An Audubon Handbook

A Chanticleer Press Edition

McGraw Hill

How to Identify Birds

John Farrand, Jr.

McGraw-Hill Book Company
New York   St. Louis   San Francisco   Auckland   Bogotá
London   Mexico   Montreal   New Delhi   Panama   São Paulo
Singapore   Sydney   Toronto

First Printing

ISBN: 0-07-019975-2

Library of Congress Cataloging-in-Publication Data
Farrand, John.
How to identify birds.
(An Audubon handbook)
Includes index.
1. Birds—Identification—Technique. I. Title.
II. Series.
QL677.F36 1988   598   87-3424
ISBN 0-07-019975-2

Prepared and produced by Chanticleer Press, Inc., New York

Color reproductions made in Italy
Printed and bound in Japan
Typeset by Dix Type Inc., Syracuse, New York

Trademark "Audubon Society" used by publisher under license from the National Audubon Society, Inc.

Jacket photographs of Great Egret by Larry R. Ditto, Harry N. Darrow, Barry W. Mansell, James F. Parnell, and Lynn M. Stone.

Design: Massimo Vignelli

The Audubon Society

The National Audubon Society is among the oldest and largest private conservation organizations in the world. With over 525,000 members and more than 500 local chapters across the country, the Society works in behalf of our natural heritage through environmental education and conservation action. It protects wildlife in more than 70 sanctuaries from coast to coast. It also operates outdoor education centers and ecology workshops, and publishes the prize-winning *Audubon* magazine, *American Birds* magazine, newsletters, films, and other educational materials. For information regarding membership in the Society, write to the National Audubon Society, 950 Third Avenue, New York, New York 10022.

About the Author

John Farrand, Jr., received a Master of Science degree in zoology from Louisiana State University. He has studied birds throughout North America and in the American tropics, Europe, and Africa. Mr. Farrand is a former editor of *American Birds*, the ornithological journal of the National Audubon Society, a past president of the Linnaean Society of New York, an elective member of the American Ornithologists' Union, and a life member of the Wilson Ornithological Society. He is co-author of *The Audubon Society Field Guide to North American Birds (Eastern Region)* and editor of *The Audubon Society Master Guide to Birding*. He has led birding tours in North America as well as in Europe, where he specializes in the natural history of Greece.

Acknowledgments

I am deeply grateful to Gudrun Buettner of Chanticleer Press for the concept of this book, which was developed through a long series of discussions. On field trips very early in my acquaintanceship with Ms. Buettner, she and her colleagues Susan Costello and Jane Opper expressed amazement at my ability to identify birds so fast. But these talented editors and close friends rightly concluded that while practice and knowledge were involved in what I was doing, there was no reason why a beginner could not quickly develop equal skill in identifying birds if told the right way to go about it.

Together, we explored the way in which bird identifications are made. We realized that the features used to identify birds—"field marks"— are easier to learn if they are grouped into categories and then examined in a particular sequence. This is exactly how experienced birders identify a bird. We began planning a book that would explain our method clearly, so that anyone could have the fun of becoming an expert birder.

I am especially indebted to Gudrun Buettner, for helping me to understand the system I was using, and for translating my abstract ideas into a practical handbook and two companion field guides. I am most grateful to Susan Costello, who has been a continuing and enthusiastic source of ideas and solutions to problems as the books have evolved; to Jane Opper, who has skillfully combined meticulous attention to details with a thorough understanding of the overall plan of the volumes, and has worked with great care along with Carol Nehring on the charts that form the basis of the book's presentation of field marks; to Carol Nehring, who has creatively organized the charts in this book and resolved design and graphic matters for all three volumes; and to David Allen for contributing his copy editing talents.

I would also like to thank Helga Lose for her expertise, perseverance, and long hours in seeing these books through their intricate production stages; Edward Douglas for his assistance in gathering the thousands of photographs that were reviewed during the preparation of these volumes; and Karel Birnbaum and Timothy Allan for their assistance in editorial matters. I am indebted once again to Massimo Vignelli for his skill in finding an effective and workable design.

Finally, I would like to express my deep appreciation to Paul Steiner for his constant enthusiasm and support during the preparation of the books in this series.

Many other colleagues and friends have encouraged me in this project. I am particularly indebted to Michael Lee Bierly, Susan Roney Drennan, Kenn Kaufman, and Kenneth C. Parkes for valuable advice. I very much appreciate the constant encouragement of Les Line, Editor-in-Chief of *Audubon* magazine.

For the remarkable collection of photographs, I am very grateful to G.C. Kelley, Robert Villani, Wayne Lankinen, and the 126 other photographers for their work, which so enhances this handbook, and for their patience and consideration.

Foreword by Les Line
To Know a Bird

Editor of Audubon *since 1966, Les Line has been the guiding spirit behind this much-acclaimed nature and conservation magazine.*

The preeminent photographer of the North American landscape and its birdlife, Eliot Porter, once summed up the secret of successful songbird portraiture: You won't find nests, he said, unless you know the birds and their habits. Know them by sight and song, of course, but more importantly, know their behavior, their habitat, their geographical distribution. "It would be folly to search for forest-adapted species in the arid Southwest, or conversely for desert types in New England. Even within one of these regions, time would be wasted looking for arboreal birds in meadows."

One of my favorite Eliot Porter photographs is of a Northern Parula flitting from its nest on a Maine island. "The parula warbler is unique among tree-nesting birds because of its nesting requirements," Dr. Porter wrote in *Birds of North America: A Personal Selection.* "Its original breeding habitat appears to have been in the region of the Gulf of Mexico, where the trailing strands of an epiphytic plant commonly known as Spanish moss, found growing on live-oak trees, is used by warblers of this species to build their nests. With the withdrawal of the glaciers, the range of these birds spread northward, and they discovered a comparable ecological niche in coniferous forests where the *Usnea* lichen or beard moss commonly grows. They use the strands of the lichen to construct pendent pouches among the clumps that festoon the spruce trees of northern forests, skillfully drawing the trailing filaments together underneath a branch to form a durable sack."

Mr. Line has birded in Japan, South Africa, Europe, Trinidad, the Galapagos, and from the Arctic Ocean to the Gulf of Mexico.

He considers himself only a "casually serious" birder with a North American life list "somewhere around 600. I stopped counting years ago."

In migration, this blue-and-gold sprite might be seen in any leafy place, including your garden or the village green. But a parula in, say, a mixed deciduous woodland in June would be highly suspect. Like the Blue Grosbeak that I spotted in my formative birding days at the edge of a forest in north-central Michigan. The habitat was generally wrong, and it was well north of the species' range. It was, however, the right place for an Indigo Bunting! But after all, I told an annoyed friend and veteran birdman, having dragged him away from the dinner table to witness my rare discovery, both are blue finches, and their songs are not that dissimilar. And, yes, the Blue Grosbeak is larger and has chestnut wing bars, but I was looking into the sun . . .

I might have been spared such ignominious moments (fortunately, no one else knew about most of my birding bungling) had this Audubon handbook, *How to Identify Birds*, and its companion field guides been available back in 1960. For John Farrand and his colleagues at Chanticleer Press have stripped away much of the veneer of mystery that has surrounded naming birds. They make it easy for the beginner to grasp a bird's "jizz"—that combination of several crucial clues that simplify identification.

Contents

Counting Hawks in Nevada

Mudflat

Sparrow-sized

Diving from Air

To master field identification,
read the chapters listed below in
sequence, from the first to the last.

Birding in Washington

Birders in North Carolina

Northern Wooded Swamp

Desert

Robin-sized

Goose-sized

Swimming with Body Low

Feeding at Flowers

Contents

Slender

Stocky

Blue

Red

Clear Whistle

Cooing

In the Mountains

By the Shore

First Steps

First Steps

There has never been a time when birds didn't fascinate me. Long before I became a serious birder, I knew the common birds around our house—robins, crows, grackles, catbirds, and sparrows, the chickadees, nuthatches, and woodpeckers that visited our feeders in winter, and a "thrush" that nested every year in a bush near our front door. On my eighth birthday, in December, I found a large owl perched in the tree only a few feet from my bedroom window—a birthday present that eclipsed all the others that year.

But seeing an owl would interest any child, and I was just as intrigued by squirrels, garter snakes, butterflies, and even the wasps that built their nests of mud or paper under the eaves of our house. It was one particular adventure that launched my lifelong interest in birds. When my parents bought a new refrigerator, they gave me the huge cardboard carton it came in. I dragged the carton out onto the lawn, sprinkled some seeds in front of it, and crawled inside to wait for birds. Some sparrows quickly arrived. As I watched breathlessly through holes in the sides of the box, they ate the seeds I had scattered. Never before had I seen wild birds so close. I can still remember one bird, so near that I could see every delicate marking. As they fed, the sparrows uttered soft notes—sounds they meant for one another and not for me—and there, crouched in the stuffy carton, I realized that these creatures had lives of their own.

A Childhood Hobby

I knew that there were many other birds in our yard and decided then and there to learn them all. That Christmas, my parents gave me a field guide. I never doubted that some day I would see every species illustrated. But how soon? A child has a lot of free time, and I spent mine learning birds. It was slow going at first, even though I thought of little else. I would find a bird and then hunt through my field guide for a picture that matched it.

It took me a while to learn how to really see a bird, and I often came home remembering features that were no help in identification. Although I made many mistakes, I gradually began to see birds more clearly. I learned that our "thrush" was actually a Brown Thrasher, that my birthday owl had been a Barred Owl, and that the sparrows that had come to my seeds were House Sparrows. Within a few months, I had found Red-breasted Nuthatches in our pine trees, and saw how they differed from the White-breasted Nuthatches that visited our feeders.

After school, in May, I would spend hours perched high in an ironwood tree by the river, waiting for migrant warblers. With my field guide in one hand and my four-power binoculars in the other, I saw Yellow Warblers, Black-throated Blue Warblers, Yellow-rumped Warblers, and once, a Cape May Warbler. As I learned how to focus on critical features of birds—their "field marks"—and to identify different species, I began to understand the features that united groups of species. I had seen the differences between the Red-breasted and White-breasted nuthatches, but I also saw that their stocky shape, large head, slender, pointed bill, and short tail are what made them both nuthatches.

Before long, my bicycle was taking me farther afield to the beaches and salt marshes on Long Island Sound. Here I found ducks, herons, sandpipers, and at dusk one cold winter day, a Snowy Owl. The list of species I had seen grew steadily longer. But it was seeing the birds and taking notes on what they were doing that I enjoyed most.

A Full-time Profession

I went away to study ornithology at college, where I learned the formal categories that scientists use to classify birds, and about the slow evolutionary processes that had produced the great variety of species. In the years that followed, I went birding in the spruce forests of northern New England and Quebec, the farmlands of upstate New York, the plains of Oklahoma and Kansas, the marshes of Louisiana, the deserts of Arizona, the prairies of Manitoba, the rocky coasts of California, and eventually overseas. Working in museums, I saw the subtle features that distinguish groups and difficult species.

Eventually I began leading field trips, both in the United States and abroad. I discovered that many beginners have trouble learning and remembering birds. Often people describe a bird to me and ask what it might have been. When I question them about the bird's bill, or shape, or behavior, I find that they don't remember these features, the "real" field marks, and as a result, they can't find the bird in their field guide and I can't identify it. This has happened to me many times, and I have often wondered why. Here were bright people with a love for nature; birding was their hobby. They were eager, armed with good field guides and decent binoculars, and often had jobs that were far more demanding than identifying birds. I decided that the problem lay not with the people, but with the little time they had to devote to their hobby. The question was how could many years of experience be used to find an easy way to teach people how to master the art of identifying birds in a reasonably short time.

An Easy, Systematic Approach

I took a careful look at how I do it. I can usually name a bird in seconds, taking in a variety of field marks all at once. It is all automatic, but I found that I was using a system. This book tells you how my system works. If you have a little patience, a love of nature and of birds, a notebook, and a reliable field guide, you, too, will soon be able to name a bird at once when you see it.

My step-by-step approach requires some practice, but no previous knowledge of birds. You can start using it the first time you go out in the field or even while looking at birds from a window. You need not worry about scientific classification, or about the definition of a songbird, or about the distinction between water birds and land birds. Instead, I deal directly with the field marks that you can see or hear the moment you find a bird.

Even if you never develop a serious interest in birds, or if you already have some birding experience, I hope this book will teach you how to look at birds and sharpen your powers of observation. A keen eye for birds will add to your enjoyment of the outdoors, whether they become a lifelong passion or just one of the many fascinating parts of the natural world around you. I wish you Good Birding.

Habitat and Size

Habitat

Since identifying a bird is a process of elimination that leads from the known to the unknown, habitat is always the first category of field marks I consider when I'm out birding. As you will learn, habitat is the most obvious characteristic of any bird. No matter where I see a bird, I always know where I am—that is, what habitat the bird and I are in.

What Is a Habitat?

Every habitat is a special environment with its own distinctive vegetation, landforms, or climate. North America offers a great variety of habitats, from suburban gardens and densely populated urban areas to dense forests, broad grasslands, deserts, alpine tundra, beaches, mudflats, freshwater marshes, inland lakes, and even open ocean. Every one of these habitats, and the others covered in this chapter, is a rewarding place to search for birds.

We find birds where we do because all bird species are adapted to making use of the food, shelter, or nesting sites that are available in certain of these habitats. Some species are so dependent on a narrow range of conditions that they are normally found in only one habitat. In fact, some birds are so highly specialized, so demanding in their requirements, that they are among North America's endangered and threatened species.

The First Step in Bird Identification

This close link between habitat and adaptations is why I find habitat so useful as a first step in sorting out birds. While species are attracted to some habitats, they avoid others; thus every habitat has its own predictable assortment of bird groups and species. On a coastal mudflat at low tide, for example, I expect to see only those birds that can find food in this distinctive saltwater environment. While almost anything is possible—I have seen Swainson's Thrushes, normally a forest-dwelling species, foraging on a mudflat on Cape Cod—I don't expect to see woodpeckers, nuthatches, warblers, flycatchers, tanagers, or sparrows, since these are normally birds of forests, open country, or residential areas. Therefore recognizing habitats and understanding their makeup is the first step in identifying any bird.

Categories of Habitats

To help you in this task, I have divided the American landscape into 23 different habitats, and have loosely grouped them into five broad categories: Salt Water, Fresh Water, Forests, Open Areas, and Urban and Residential Areas. Some ornithologists recognize dozens of additional habitats, based on narrower distinctions between dominant plants, geology, temperature, or rainfall. I find that the 23 habitats described in this chapter are a manageable number, cover the full range of variation, and are distinct enough to be useful in the process of elimination. Moreover, you need not be a biologist, because identifying these habitats requires little or no knowledge of botany. To assist in recognizing a habitat and determining its location, a map has been prepared for each of the five major categories. The maps show all of North America except the Arctic. Where possible, these

Habitats
■ Salt Water
■ Fresh Water
■ Forests
Open Areas
■ Urban and Residential Areas

maps indicate the distribution of most of the habitats included in the larger category. Each map is followed by a photographic essay that illustrates and describes each of the various habitats, with between one and sixteen photographs. Finally, on the charts at the end of this chapter, I have listed all North American bird groups, as well as very striking species, according to the habitats in which they are normally found.

How to Recognize a Habitat

Study these pages carefully, look at the photographs and maps, and soon you will have no trouble deciding instantly what habitat you are in. You will be able to distinguish between the different types of forests—for example, an Eastern Deciduous Forest, with its tall oaks, birches, maples, beeches and other broad-leaved trees, and a Boreal Forest, with its predominance of spruces and firs; or between beaches and mudflats, two shoreline habitats with very different types of food and very different birds.

In learning to identify habitats, your notebook can be very useful. Whenever you go into the field, and even on your first walks in search of birds in your backyard, take a moment to jot down a description of the habitat. Ask yourself questions. Are there trees in the habitat? If so, are they evergreens or broad-leaved trees? Is the ground bare, or is it covered with grass, dried leaves, or dense brush? Are there thick shrubs for birds to hide in? Are you in the mountains or in a lowland? At the shore, is the tide out, so that mudflats are exposed, or is it in, leaving only beaches as resting places for birds? Taking notes will remind you of details that might be helpful later, and what is equally important, taking notes will cause you to become more observant. It is a curious fact that the act of taking notes itself, as a safeguard against forgetting our observations, often plants them firmly in mind, assuring that they won't be forgotten.

The charts at the end of this chapter will tell you what birds are found in the habitat you are visiting. The companion field guides will indicate birds in your geographical area. Once you know what habitat you are in and have learned what species and groups are likely to be seen there, you will find that you have eliminated a great many birds. Instead of all the hundreds of species in your field guide, you will have narrowed the possibilities to only a few dozen. The task of identification has already become greatly simplified.

One word of caution: You may have heard birders talking about water birds and land birds. You might expect all birds in the water habitats to be water birds, and all birds in the other habitats to be land birds. Unfortunately, this is not so. The mere presence of a bird at water does not make it a water bird, because land birds frequently visit water to bathe or drink, and can be found in almost any saltwater or freshwater habitat. Furthermore, the special adaptations of water birds are sometimes impossible to see, and some water birds may turn up far from water. This traditional distinction, then, will not help you identify birds, and so I have chosen size as the field mark you should focus on after you have learned about habitat.

Salt Water

Bounded on the east by the Atlantic Ocean, on the south by the Gulf of Mexico, and on the west by the Pacific, North America offers birds rich and varied saltwater habitats. Besides the oceans, three habitats can be seen here, but Inshore Waters, and the Mudflats and Tidal Shallows that border them, are not mapped because they are too intricate and often overlap.

Salt Marshes cover large areas on the Atlantic and Gulf coasts, but on the Pacific Coast they are too limited in extent to appear on a map of this scale; they are found chiefly near San Diego and on San Francisco and Tomales bays in California, and on Puget Sound in Washington.

Habitats

Open Ocean

Inshore Waters

Rocky Shores and Coastal Cliffs

Beaches and Coastal Dunes

Mudflats and Tidal Shallows

Salt Marshes

Open Ocean

The Open Ocean habitat consists of all of the surface of the sea that is out of sight of land. It is accessible only by boat.

Twelve miles offshore, Gulf of Mexico

The few groups of birds that have successfully adapted to this demanding environment are easy to recognize. Only a small number of species are likely to be found in any given area of ocean.

Inshore Waters

Inshore Waters include not only the ocean just beyond the surf line, but also bays, harbors, sounds, estuaries, tidal inlets, and other large and relatively sheltered bodies of salt water along the coast.

Surf line, Monterey Peninsula, California

Whooping Cranes on bay, Aransas, Texas

These habitats offer a greater
variety of food resources than is
available farther out at sea, so that
a larger number of bird groups
may be found here, including
several that also feed over the
Open Ocean.

Tidal inlet, Acadia National Park, Maine

Great South Bay, Long Island, New York

Rocky Shores and Coastal Cliffs

Rocky Shores—with their tide pools, seaweed-covered boulders, and other feeding sites, as well as convenient perches and inaccessible nesting ledges—attract birds in search of food or safe breeding areas.

Cliffs, California

Common Murre, Newfoundland

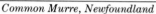

Western Gull, California

Rocky shore, British Columbia

Some of the same birds found in
Inshore Waters may be swimming
in the nearby surf.

Rocky shore, Oregon

Northern Gannet, Newfoundland

Atlantic Puffin, Maine

Black-legged Kittiwake, Nfld.

Beaches and Coastal Dunes

Beaches are mainly feeding and resting places for birds, although a few species nest on the sand above the high-tide line. Different birds inhabit the adjacent Coastal Dunes, but all of them make frequent visits to the beach.

Snowy Owl on dune, Long Island, New York

Black-bellied Plovers on beach, Massachusetts

Birds found in Inshore Waters may also be seen from the beach swimming on the water.

Coastal dunes, Texas

Ruddy Turnstones and Laughing Gulls on beach, New Jersey

Mudflats and Tidal Shallows

Bordering sheltered Inshore Waters, this habitat consists of mudflats and nearby shallows. It is at its best at low tide, when many birds are attracted by the rich food supply here.

Mudflats along Washington coast

Shorebirds in tidal shallows, New Jersey

At high tide, the birds move to beaches or rocks, or out into deeper water, where they rest, waiting for the tide to recede so they can begin foraging again.

Salt Marshes

Salt Marshes are flat expanses of marsh grasses and rushes, periodically inundated by salt water and often broken up by a network of tidal creeks and ditches.

Tidal creek, Acadia National Park, Maine

Canada Geese over salt marsh, Maryland

Most of the birds of Salt Marshes
are water birds, but some are
members of grassland groups,
adapted to this special marine
environment.

Tidal creeks, California

Salt marsh, New York

Fresh Water

Freshwater Marshes have the largest and most varied birdlife of any freshwater habitat, because they offer an abundance of plant and animal food, as well as safe nesting sites in the dense marsh vegetation. Lakes, Reservoirs, Ponds, and Rivers, and Streams and Brooks also have their own distinctive groups of birds. All of these habitats are distributed throughout North America.

Southern Wooded Swamps, with their mixture of water birds and birds of the forest, are scattered across all the southeastern states. Although the Mangroves of southern Florida grow in salt or brackish water, their birds are the same as those of other Southern Wooded Swamps, so these two habitats are treated as one.

The map shows North America's most important lakes and rivers.

Habitats
Freshwater Marshes
Lakes, Reservoirs, Ponds, and Rivers
Streams and Brooks
Southern Wooded Swamps and Mangroves

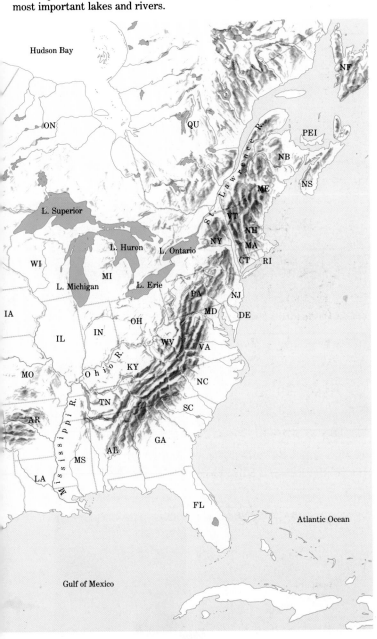

Freshwater Marshes

Freshwater Marshes are open expanses of fresh water with marsh grasses, cattails, tules, reeds, water lilies, and other aquatic plants. Although these marshes may have brushy borders or a few isolated dead snags, they are essentially treeless.

Malheur, Oregon

Point Reyes, California

Tules, California

Cattails, New York

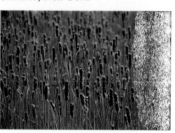

Yellow-headed Blackbird, Wyoming

Least Grebe, Texas

Northern Jacana, Texas

Common Moorhen, Florida

The best time to visit one of these
marshes is at dawn, when the
birds are most active and noisy.

Canaan Valley, West Virginia

Shelter Island, New York

Water lilies, Maine

Acadia National Park, Maine

Eared Grebe, Oregon

Red-necked Grebe, Minnesota

Least Bittern, Alabama

American Bittern, Nebraska

Lakes, Reservoirs, Ponds, and Rivers

Lakes, ponds, and large, slow-moving rivers are open, quiet inland waters much visited by birds that feed on fish, crustaceans, insects, and aquatic plants. Some birds, such as gulls, ducks, and geese, also use these bodies of water as safe places to rest and preen.

Snake River, Idaho

Saint Mary Lake, Glacier National Park, Montana

Walden Pond, Concord, Massachusetts

Missouri River, Montana

Raptors, swifts, and swallows, birds usually found in other habitats, sometimes forage over open water.

Pond, near Mount Bigelow, Maine

Pond, Rocky Mountains, Wyoming

Lake, Adirondacks, New York

Pond, Bombay Hook Island, Delaware

Streams and Brooks

Small, shallow rivers, streams, and brooks, often flowing swiftly over stony beds, wind across the countryside, through habitats whose birds may come to running water to bathe or drink. But there are birds whose lives are dependent on streams, creeks, and brooks, and that never or only rarely leave them to enter an adjacent forest or grassland.

Stream, Idaho

Stream, Vermont

Brook, New Jersey

Brook, California

Only these truly stream-dwelling
birds are considered inhabitants of
the Streams and Brooks habitat.

Stream, Montana

Brook, Montana

Stream, Michigan

Stream, North Carolina

Southern Wooded Swamps and Mangroves

This habitat includes cypress swamps, hardwood swamps along rivers and lake margins, bogs, and the hardwood hammocks and coastal mangroves of Florida.

Cypress-gum swamp, N.C.

River swamp, Louisiana

Great Egret, Florida

Yellow-crowned Night-Heron, N.C.

Combining both birds of the forest and water birds, Southern Wooded Swamps are among North America's richest bird habitats.

Palmetto swamp, South Carolina

Egret colony, Louisiana

Wood Duck, Virginia

Limpkin, Florida

Forests

Forests are complex habitats that include the leaf litter and herbs of the forest floor, the trunks of trees, one or more shrub layers, and the leafy canopy formed by the tallest trees. These different environments offer a variety of niches for many species of birds, so that forests—even the cold Boreal Forest—are among the richest habitats for birds.

Northern Wooded Swamps are scattered across southern Canada and the north and central United States, while Thickets can be found throughout the continent. Both habitats are too local to be shown on a map of North America. The others are mapped here.

Habitats
- Boreal Forest
- Eastern Deciduous Forest
- Northern Wooded Swamps
- Southeastern Pine Forest and Pine Barrens
- Western Coniferous Forests
- Dry Western Woodlands, Streamside Forest, and Chaparral
- Thickets

Boreal Forest

The vast Boreal Forest, or *taiga*, is composed of spruce, fir, and larch. It extends from Alaska, northern Quebec, and Newfoundland south to the northern boundary of the United States, and in the mountains as far south as Georgia. This habitat is also found at high elevations in the mountains of the West.

Spruce and Balsam Fir, Maine

Spruce, Maine

With its cold winters and limited numbers of plant species, the Boreal Forest has fewer bird species than other forest types, but many of these species are found in no other habitat.

White Spruce, New Hampshire

Black Spruce, Newfoundland

Eastern Deciduous Forest

Much of the eastern United States is covered by a forest of oak, beech, maple, birch, hickory, and other broad-leaved trees that lose their leaves in winter. This area represents the original forest, logged off by the earliest European settlers, but now reappearing where farmland has been abandoned.

Indigo Bunting, Texas

Sugar Maples, New York

Mixed forest, Vermont

Eastern Screech-Owl, Connecticut

Eastern Deciduous Forest includes mixed forest—northern areas where pines and hemlocks mingle with the deciduous trees—and comprises more kinds of trees than any other North American forest. It also undergoes great changes from one season to the next. A large number of birds have adapted to this richly varied forest environment.

Yellow-bellied Sapsucker, Ohio

Maples and birches, Vermont

Beech forest, Michigan

Red-bellied Woodpecker, Michigan

Northern Wooded Swamps

Scattered through the forests of Canada and the northern United States are naturally flooded areas: maple or alder swamps, river-bottom forests, and bogs. These wooded swamps are a favored habitat for several land birds.

Black Spruce bog, Michigan

Spruce-fir bog, New York

Red Maple swamp, New York

Cedar bog, Minnesota

If you see water birds, check the appropriate nearby freshwater habitat.

Rhododendron swamp, Virginia

Spruce-larch bog, Minnesota

Red Maple swamp, Massachusetts

Spruce-fir bog, New York

Southeastern Pine Forest and Pine Barrens

An extensive pine forest stretches from Delaware southward on the coastal plain to Florida and west to Texas. It varies from open stands of tall, straight trees to dense pine scrub. Broad-leaved trees grow along streams and in disturbed areas. The Pine Barrens of New Jersey, New York, and Michigan are also part of this habitat.

Slash Pine, Florida

Pine forest, Maryland

Pinelands often may appear
monotonous, an unpromising place
for birds, but a surprising number
of species can be found here, and a
few, like the Red-cockaded
Woodpecker and Brown-headed
Nuthatch, are found in no other
habitat. The rare Kirtland's
Warbler nests only in a small area
of pine barrens in Michigan.

Chipping Sparrow, New York

Red-bellied Woodpecker, North Carolina

Western Coniferous Forests

The coniferous forests of the West grow at many different elevations, each with its own distinctive trees and other plants. From the moss-covered Olympic forest of the humid Pacific Northwest, and the dense forests of pine and spruce on steep mountain slopes, to the forest of stunted pines and firs at timberline, each of these forest types has its own distinctive group of bird species.

Engelmann Spruce, Montana

Bald Eagle, British Columbia

Lodgepole Pine, Montana

Olympic forest, Washington

Often a bird found at one elevation is replaced at higher or lower elevations by closely related species. The drier, more open woodlands of pine and oak, or of pinyon and juniper in California and the Southwest, are considered a different habitat, Dry Western Woodlands.

Broad-tailed Hummingbird, Colo.

Steller's Jay, Colorado

Cedar and hemlock, Montana

Jeffrey Pine, California

Dry Western Woodlands, Streamside Forest, and Chaparral

Dry, open woodlands of oak or pinyon pine and juniper are found on the lower slopes of the western mountains, between the coniferous forest and the dry lowlands. Streamside forest shelters birds not found in nearby deserts or grasslands.

Pinyon-juniper, California

Oak woodland, California

Whiskered Screech-Owl, Arizona

Summer Tanager, Arizona

Chaparral is a dense growth of
thorny shrubs and small trees that
covers hillsides in both California
and the Southwest; birds here are
more often heard than seen.

Streamside forest, California

Chaparral with pine, California

Vermilion Flycatcher, Texas

White-crowned Sparrow, Arizona

Thickets

A thicket is a dense stand of shrubs, small trees, or weeds, thicker than brushy Open Country, but not as tall as a forest or woodland. Thickets are often found in low, moist spots, along streams, or at the edge of forests. Thickets offer both shelter and an abundant source of food—insects and fruit—for many birds.

Willow thicket, California

Willow thicket, Utah

Because ornamental plantings often resemble natural thickets, a number of species that occur in this habitat are also found in Residential Areas.

Thicket, Michigan

Thicket, California

Open Areas

Open areas include all habitats on dry land without an extensive growth of trees. Open Country and Grasslands comprise both the grassy plains and prairies of the western and central parts of North America, as well as man-made grasslands. Scattered Groves are found wherever these open lands occur. Barren desert areas, where even grass is lacking, are also considered Open Country.

Many of the birds of open habitats are easy to see and are often boldly patterned, but others skulk in the grass and are patterned in browns and grays.

Because they are so widespread, Open Country, Grasslands, and Groves cannot be shown on a map. Other open habitats are more restricted and are mapped here.

Habitats

Open Country, Grasslands, and Groves

Deserts and Sagebrush Plains

Mountain Cliffs, Gorges, Mesas, and Bare Rocky Slopes

Alpine Tundra and Meadows

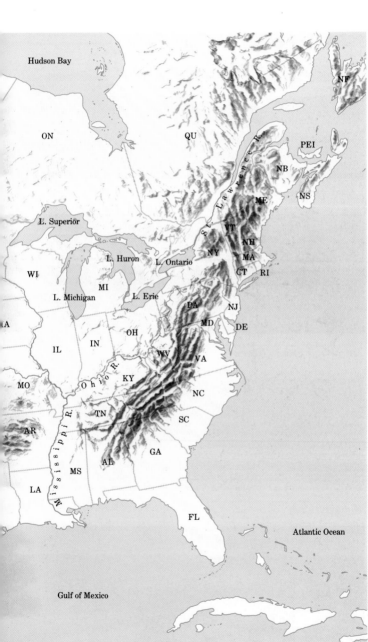

Open Country, Grasslands, and Groves

This habitat includes not only natural grasslands and prairies, but also man-made grasslands and cultivated areas such as pastures, fields, meadows, croplands, airports, golf courses, and the grassy margins of roads. Brushy Open Country, overgrown with weeds or low bushes, as well as areas with no vegetation at all, such as barren deserts and plowed fields, are part of this habitat.

Pastures, Connecticut

Bobolink in pasture, Pennsylvania

Mesquite grassland, Texas

Killdeer, Texas

These open lands sometimes contain isolated groves of trees that offer shelter for open-country birds and for some species that are otherwise restricted to forest habitats.

Grassland with grove, South Dakota

Snow Geese in cropland, New Mexico

Coastal prairie, Texas

Canada Geese in cornfield, Maryland

Deserts and Sagebrush Plains

Shrubs, yucca, and agave, as well as cholla, prickly pear, and other cacti of the southwestern deserts, provide habitat for a surprising number of birds. To the north, a smaller but equally distinctive group of birds may be found in the arid Sagebrush Plains of the Great Basin.

Yuccas, Arizona

Agaves, California

Cactus Wren, Arizona

Scaled Quail, Arizona

Cholla, California

Saguaros, Arizona

Mourning Dove, Arizona

White-winged Dove, Arizona

Barren desert areas with no
vegetation are considered Open
Country.

Desert, Arizona

Sagebrush plain, California

Costa's Hummingbird, Arizona

Desert dunes, Arizona

Sagebrush, California

Sagebrush plain, Wyoming

Pyrrhuloxia, Texas

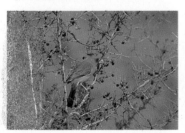

Wild Turkey, Wyoming

Mountain Cliffs, Gorges, Mesas, and Bare Rocky Slopes

Much of the landscape of the West consists of rocky cliffs, canyon walls, and boulder-strewn slopes that support little or no plant life. These seemingly inhospitable places are the preferred habitat of several birds, some of them found nowhere else.

Mesas, Utah

Chukar, Nevada

Bald Eagle, British Columbia

Mountain cliffs, Maine

These birds are often attracted by nesting sites that are inaccessible to predators, or wander into these barren places in search of food.

Bare rocky slope, Utah

American Kestrel, B.C.

Canyon Wren, Arizona

Canyon, Utah

Alpine Tundra and Meadows

Above treeline in the mountains of the West are grassy meadows with wildflowers and patches of tundra with scattered willow thickets.

Alpine meadow, Washington

Alpine meadow, California

The birdlife of these high mountain
habitats is a mixture of species
that also live in arctic tundra and
birds that have moved upward
from the nearby forested slopes in
search of food.

Alpine meadow, Washington

Alpine meadow with distant tundra, Montana

Urban and Residential Areas

These man-made habitats have expanded rapidly in the twentieth century. Both may appear to lack the resources that birds need, but there are few predators in cities and suburbs, so that safe nesting and roosting sites are abundant.

People provide food both in bird feeders and accidentally in the form of spilled grain and discarded bread. Artificial plantings often resemble the plant life of forests and thickets, and attract many birds from these nearby habitats.

Habitats
Urban Areas
Residential Areas and Parks

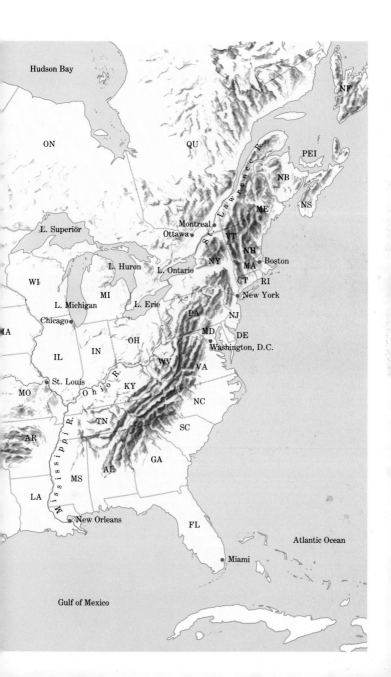

Urban Areas

The heavily built-up parts of cities, where all you can see is concrete, steel, and glass, may seem to be a poor place to look for birds. But many species have adapted to this vast world of skyscrapers and apartment houses.

Glaucous-winged Gull, Seattle

Manicured city parks are not part
of this habitat, but are included in
Residential Areas and Parks. For
wetlands, check the appropriate
salt- or freshwater habitat.

Residential Areas and Parks

Millions of acres of North America are covered by buildings and houses with lawns, shade trees, and shrubbery, or by gardens and manicured parks. This is a new bird habitat, but many species have adjusted to it, and a residential area is an excellent place to find birds.

Suburban lawn, Seattle

Broad-tailed Hummingbird, Colo.

Tree Swallow, Colorado

American Robin, New York

Open water and wetlands are not
included in this habitat, but belong
to a salt- or freshwater habitat.

Park, Portland, Oregon

Townsend's Warbler, Oregon

Chipping Sparrow, New Jersey

Rufous Hummingbird, California

Size

The 23 habitats you have just learned to identify are the home of 65 bird families, consisting of as few as a single species or as many as 130. In all there are about 650 species that regularly feed or nest in these 23 habitats of our continent. Birds in these 65 families may differ greatly in size, shape, and behavior from members of other families. Seeing and noting these differences will lead you to positive and accurate identifications of both bird groups and species.

What Is Size?

In the many years I have been identifying birds, I have found that once I know a bird's habitat, the best feature to focus on next is its size. While habitat is the most obvious field mark of any bird, size is another feature we cannot help seeing, no matter how briefly we glimpse a bird.

When I refer to size, I mean the size of a bird's body, and not its overall length or its shape. Many birds have long bills, necks, tails, or legs, all features that tend to influence one's impression of a bird's size. Considering the overall length of a bird by including a long neck or tail would make it difficult to group birds by size, because we would be dealing with individual, subjective impressions rather than with facts. The same applies to a bird's shape. Some birds may be slender or stocky: a large head, short neck, or short tail can make a bird look stocky; and a long neck or long tail can make a bird look slender. Any of these variations can influence our perception of a bird's size. Stocky birds, for example, may appear larger than slender ones of the same size, so I caution you to note the length of the body alone when assessing a bird's size. This takes practice.

Seven Size Categories

I have found that the easiest way to gauge the size of a bird is to compare it to a species one already knows. For this reason, I have grouped all birds into seven size categories. Five of these are based on common and widespread birds that most of us already know and can recognize easily. From smallest to largest, they are the House Sparrow, the American Robin, the "pigeon" or Rock Dove, the American Crow, and a typical goose, such as most Canada Geese or the Snow Goose. Two additional categories represent the extremes—the very small birds like hummingbirds at the low end of the scale, and the very large birds like pelicans at the other end.

On the following 28 pages we illustrate these seven size categories. In each category you will find the yardstick species, such as the House Sparrow, and several other birds that belong to the same size group. On each page we have included a "yardstick" that is divided into seven sections, each one of which represents one of my seven size categories. We have also used this yardstick in the companion field guides to help indicate the size of each species illustrated and described. As you will see on the following pages, the birds in this section have been arranged from the smallest to the largest, and each group ends on a page facing the next larger size category. The yardsticks will make it easier to compare sizes as you use this section. Take time to study the photographs. As you compare the birds in a

Yardsticks

	Very Small
	Sparrow-sized
	Robin-sized
	Pigeon-sized
	Crow-sized
	Goose-sized
	Very Large

size category, note their different shapes, and features such as bills, legs, and tails. Within each group, all the birds are illustrated in their proper size relationship. You can see that some birds, like the Red-breasted Nuthatch and Bushtit, are small enough to be placed in the same category as most of the hummingbirds, and that many gulls are pigeon-sized or crow-sized, but some gulls—including the widespread Herring Gull—are large enough to be considered goose-sized. Within each of the seven size categories, there is a size range—not all birds in the category have exactly the same body length.

How to Judge Size

Once you have the seven size groups clearly in mind and you feel comfortable with the idea of assigning birds to size categories, go out into the field and practice judging the sizes of birds. It is best to start by finding one of the yardstick birds—a House Sparrow or American Robin. Using your notebook, jot down which species it is and then compare the bird with objects around it. Note how large or small it is in relation to nearby plants, or if it is in a tree, compare it to the size of surrounding leaves. If you are in your backyard, compare its size with that of shrubs, flagstones, or any other objects you are used to seeing. By making these comparisons and writing them down, you will quickly become familiar with the size of the yardstick birds, and of any other birds you find.

Remember that special circumstances can influence your perception of a bird's size. When you see a bird in a situation where there are no other objects for comparison, it often seems larger than it really is. For this reason, fog can make birds seem larger than life, and shorebirds on a wide, featureless mudflat may look robin-sized when they are actually no larger than sparrows. At the other extreme, birds that are surrounded by large objects tend to look smaller than they really are. Seabirds can be dwarfed by large waves on the ocean, and small songbirds, such as warblers and vireos, can seem truly tiny as they forage in trees with large leaves.

I recommend that you make a point of designating your own yardstick birds in any habitat you visit frequently. They should be species that you have learned are common in the habitat, so that you can count on them being there to serve for comparison. Once you have built up a list of such personal yardstick birds, the value of notes you take on an unfamiliar bird will be greatly increased if you can write down something like "About the size of a meadowlark" or, at the beach, "About the size of a Sanderling."

Very Small

Here are the smallest of all our birds, including most of the hummingbirds and some tiny songbirds, among them the kinglets, gnatcatchers, and the smallest warblers.
All are land birds found mainly in forests, open country, and residential areas.

Bushtit

Brown-headed Nuthatch

Very small birds are generally
quite active, foraging rapidly in
foliage, and have rapid wingbeats.
Most birds in this group have
heads that look large in proportion
to the body.

Rufous Hummingbird

Sedge Wren

Very Small

Lesser Goldfinch

Red-breasted Nuthatch

Sparrow-sized

The sparrow-sized birds are a varied group that includes a host of typical songbirds, as well as the swifts and swallows, the largest hummingbirds, some diminutive doves, owls, and woodpeckers, and the very smallest water birds.

Black-capped Chickadee

White-breasted Nuthatch

Sparrow-sized

Many of the birds in this group have the same shape as the House Sparrow, but others may be differently built. Don't let a large head or a long tail fool you into thinking a bird is bigger than it is.

///////

House Sparrow

Semipalmated Plover

For example, the Semipalmated
Plover has a large head, like all of
the plovers, but is only sparrow-
sized.

Say's Phoebe

Yellow-rumped Warbler

Sparrow-sized

Inca Dove

American Dipper

Robin-sized

Songbirds and shorebirds form the majority of this size category, which also includes some chickenlike birds, such as quails, some small owls, and our smallest raptor, the American Kestrel, as well as the littlest gulls and terns.

Mourning Dove

Cactus Wren

Robin-sized

Hairy Woodpecker

American Robin

Despite their stocky build, the small owls are really no bigger than the American Robin. Most of the woodpeckers are also robin-sized.

Northern Mockingbird

Burrowing Owl

Robin-sized

Western Meadowlark

Long-billed Dowitcher

Although many of the songbirds have the same shape as the American Robin, shorebirds and other water birds included here clearly have their own unique build. In judging the size of the Long-billed Dowitcher, don't be misled by its long bill or long legs.

Red-winged Blackbird

Killdeer

Robin-sized

Eastern Kingbird

Northern Cardinal

Pigeon-sized

Our common "pigeon"—the Rock Dove—is the yardstick bird for this size category. In addition to some other doves, this group also includes a few large songbirds, some small raptors, and several kinds of water birds.

Forster's Tern

Clark's Nutcracker

Pigeon-sized

Among the pigeon-sized water birds are some gulls, terns, grebes, the smallest of the ducks, and the Least Bittern, a tiny marsh-dwelling member of the long-legged wader group.

Sharp-shinned Hawk

Greater Yellowlegs

When judging the size of water birds, be sure to take into account the surroundings—a large expanse of water or large waves can make birds look smaller than they are.

Least Bittern

Eared Grebe

Pigeon-sized

▨▨▨▨▨▨▨░░░░░

Rock Dove ("Pigeon")

Black Guillemot

Crow-sized

The crows and ravens, largest of
the songbirds, are in this size
category, along with many ducks
and other swimming birds, the
largest shorebirds, and some long-
legged waders.

Laughing Gull

Blue-winged Teal

Crow-sized

Several chickenlike birds, raptors, and owls, as well as our very largest woodpecker, the Pileated Woodpecker, are crow-sized. Always compare the size of the bird you have seen with a mental image of the yardstick bird.

Red-tailed Hawk

American Crow

With practice, judging size will
become almost automatic and
greatly aid in identification.

Short-eared Owl

Ruffed Grouse

Crow-sized

Spruce Grouse

Wood Duck

Goose-sized

The majority of birds in this group are water birds, including most geese, the largest ducks, some loons, some cormorants, several of the long-legged waders, and the largest gulls.

White Ibis

Snow Goose

Goose-sized

The long-legged waders are easily recognized by their long legs and long necks. But do not be tricked by these extra-long features; these birds are actually no larger than a goose.

▨▨▨▨▨▨▨▨▨▨▨▨▨

Anhinga

Tricolored Heron

Among the few land birds in this
size category are owls, raptors,
and chickenlike birds.

Little Blue Heron

Roseate Spoonbill

Goose-sized

Herring Gull

Turkey Vulture

Very Large

This group of North American giants includes both water birds and land birds. The large size of many of these birds can be determined easily, even at a great distance, by their slow and often ponderous wingbeats.

Great Cormorant

Wild Turkey

Very Large

Many of the very large birds are water birds, including long-legged waders like the cranes, the Wood Stork, and the "Great White Heron." Also included here are the swans, the largest geese, loons, cormorants, and boobies, the Northern Gannet, the albatrosses, and the pelicans.

"Great White Heron"

Bald Eagle

Land birds are represented by the
largest raptors—eagles and the
California Condor—as well as the
Wild Turkey.

Brown Pelican

Tundra Swan

How to Use the Habitat and Size Charts

Salt Water
- ■ Open Ocean
- ▨ Inshore Waters
- ■ Rocky Shores, Coastal Cliffs
- ▨ Beaches, Coastal Dunes
- ■ Mudflats, Tidal Shallows
- ▨ Salt Marshes

Fresh Water
- ■ Freshwater Marshes
- ▨ Lakes, Ponds, Rivers•
- ■ Streams, Brooks
- ■ Southern Wooded Swamps•

Habitat and size are the first two field marks you can use to begin identifying birds. As I mentioned earlier, bird identification is a process of elimination—you narrow down the number of birds until you reach the species you have seen. Becoming familiar with habitats and getting to know the various kinds of birds found in any particular habitat takes practice. It also takes a little work to learn to categorize birds by size. To help you relate habitat and size to birds, you need a tool; so, in the charts on the following pages I have organized all the birds covered in the accompanying field guides according to habitat and size.

Habitat
This book recognizes 23 bird habitats, which are listed across the top of each chart, beginning with Open Ocean and ending with Residential Areas. As shown above, these habitats are color-coded for easy reference. The names of some of the habitats are abbreviated; a dot next to the name of a habitat indicates such an abbreviation. If you do not remember the full name of one of these habitats, refer back to the habitat descriptions.

Size
You will recall that we have arranged all birds into seven size categories: Very Small, Sparrow-sized, Robin-sized, Pigeon-sized, Crow-sized, Goose-sized, and Very Large. Now that you have practiced categorizing common birds in your area by size, you can use the Habitat and Size Charts. On the left side of each chart the birds are listed alphabetically according to their respective sizes.

The Bird Groups
Because of the large number of birds, these two field marks—habitat and size—will help you narrow your choice down to bird groups, for the most part, rather than to individual species. These groups do not always represent families and subfamilies, the formal bird categories used by ornithologists. Instead, I have arranged the species in more informal groups whose members are very similar to one another, and whose resemblance can be seen clearly by a beginner. For example, orioles are members of the blackbird family, but because of their bright colors they are easily distinguished from typical blackbirds and are listed separately. Refer to the section Bird Groups for a description of my 62 groups; to learn the species included in unfamiliar or large groups, such as chickenlike birds, shorebirds, long-legged waders, or raptors, check the Group Index.

In addition to groups, I am also listing many single species. These species include some that are so distinctive in behavior, shape, or color that they form a "group" of one, such as the Anhinga. Other birds listed separately are those likely to be familiar to most readers already, such as the American Robin, or very distinctive species like the Purple Gallinule.

Although I have not divided the birds into water birds and land birds, I feel that you should become familiar with the two terms that are used collectively by birders. I have therefore placed a blue dot in front of a group or species name to indicate that it belongs to the water bird group; birds that do not have a dot are land birds.

Forests	Open Areas
■ Boreal Forest	Open Country•
■ Eastern Deciduous Forest	Deserts, Sagebrush Plains
■ Northern Wooded Swamps	■ Mountain Cliffs, Rocky Slopes•
■ Southeastern Pine Forest•	■ Alpine Tundra and Meadows
■ Western Coniferous Forest	**Urban and Residential Areas**
■ Dry Western Woodlands•	■ Urban Areas
Thickets	■ Residential•

Following the name of a bird group, you will find in parentheses the number of species in that group for that particular size category. Some groups include only a few species, while others contain more than 40. When there is only one species indicated—for example, "Flycatchers (1)" in the Very Small group—this means that there are other flycatchers, in other size categories, in this instance, in Sparrow-sized and Robin-sized. Note also that the absence of a number signifies a single distinctive species.

Using the Charts

You may want to study the charts first before you go into the field. Refer to the habitat column for the habitat you are planning to visit, and check which groups of birds you will encounter. While group members always share common characteristics, such as similar shape (long-legged waders), behavior (woodpeckers), or color (blackbirds), members of a group can vary in size. Among these birds, size is not only important in distinguishing groups, but also in identifying species. Select a group of birds you may want to look for. Try one with only a few members, such as the two kinglets (in the Very Small category). Look up this group in the field guides. You will discover that the two kinglets are the Golden-crowned and the Ruby-crowned. Then if you are visiting a Boreal Forest, you may expect to see kinglets among other birds found in this habitat. You should also note that kinglets may possibly be seen in as many as eight habitats. The charts may be used in another way. If you are on the Open Ocean, for example, and you see a Sparrow-sized bird, it can only be one of two storm-petrels, because they are the only birds of this size in this habitat. Then refer to your field guide to determine which species you saw. This exercise will help you to get to know the many different birds or bird groups that inhabit our continent. Soon their names will become part of your everyday vocabulary.

Mastering Habitat and Size

Use your notebook in the field to assign birds to their proper size category. And, of course, do not forget to write down the habitat. Then, on the spot or later at home, check the charts under the bird groups listed in the size category you have seen. You may recognize the group and then turn to the companion field guide for verification. Be sure to read the range statement to see if the species is in your area. Chances are, using only two field marks, habitat and size, you cannot decide on a group or species and thus you will need to know more. You may see a very small bird in a thicket, but you will be unsure to which of five groups it belongs. Having made the best use of habitat and size, you are now ready to turn to other field marks that can pinpoint groups and species. It is time to move on to the next four field marks—behavior; shape and posture; color and pattern; and voice.

Habitat and Size

● *Habitat Abbreviated*

	Open Ocean	Inshore Waters	Rocky Shores, Coastal Cliffs	Beaches, Coastal Dunes	Mudflats, Tidal Shallows	Salt Marshes	Freshwater Marshes
Very Small							
Bushtit							
Finches (1)							
Flycatchers (1)							
Gnatcatchers (2)							
Hummingbirds (13)							
Kinglets (2)							
Nuthatches (3)							
Verdin							
Vireos (1)							
Warblers (2)							
Wrens (2)							■
Sparrow-sized							
Blackbirds (2)						■	■
Bluebirds (3)							
Budgerigar							
Bulbul, Red-whiskered							
Buntings (7)			■				
Chickadees (4)							
Creeper, Brown							
Dickcissel							
Dipper, American							
Doves (2)							
Finches (11)							
Flycatchers (16)							
Grosbeaks (1)							
Hummingbirds (2)							
Juncos (2)							
Larks (2)				■			
Longspurs (4)				■			
Nuthatches (1)							
Orioles (1)							
Owls (3)							

Lakes, Ponds, Rivers•	Streams, Brooks•	Southern Wooded Swamps•	Boreal Forest	Eastern Deciduous Forest	Northern Wooded Swamps	Southeastern Pine Forest•	Western Coniferous Forest	Dry Western Forest	Western Woodlands•	Thickets	Open Country•	Deserts, Sagebrush Plains	Mountain Cliffs, Rocky Slopes•	Alpine Tundra and Meadows	Urban Areas	Residential•

Habitat and Size

- *Habitat Abbreviated*
- Water Birds

	Open Ocean	Inshore Waters	Rocky Shores, Coastal Cliffs	Beaches, Coastal Dunes	Mudflats, Tidal Shallows	Salt Marshes	Freshwater Marshes
Sparrow-sized continued							
Pipits (2)							
● Rails (2)						□	■
● Shorebirds (9)				□	■	□	■
Sparrows (31)				□		□	■
● Storm-petrels (1)	■	□					
Swallows (6)						□	■
Swifts (2)			■				
Tanagers (1)							
Thrushes (4)							
Titmice (3)							
Towhees (1)							
Vireos (10)							
Warblers (43)				□			■
Waterthrushes (2)							
Waxwings (1)							
Woodpeckers (2)							
Wrens (6)						□	■
Wrentit							
Robin-sized							
● Auks (4)	■	□	■				
Blackbirds (6)						□	■
Cardinal, Northern							
Catbird, Gray							
Chickenlike Birds (6)							
Cuckoos (2)							
Doves (5)							
Flycatchers (14)							
● Grebes (1)							■
Grosbeaks (5)							
● Gulls (1)		□					■
● Jacana, Northern							■
Jays (6)							

Lakes, Ponds, Rivers•
Streams, Brooks•
Southern Wooded Swamps•
Boreal Forest
Eastern Deciduous Forest
Northern Wooded Swamps
Southeastern Pine Forest•
Western Coniferous Forest
Dry Western Forest
Thickets
Open Country•
Deserts, Sagebrush Plains
Mountain Cliffs, Rocky Slopes•
Alpine Tundra and Meadows
Urban Areas•
Residential•

Habitat and Size

- *Habitat Abbreviated*
- Water Birds

	Open Ocean	Inshore Waters	Rocky Shores, Coastal Cliffs	Beaches, Coastal Dunes	Mudflats, Tidal Shallows	Salt Marshes	Freshwater Marshes
Robin-sized continued							
Kingfishers (1)							
Meadowlarks (2)							
Mockingbird, Northern							
Myna, Crested							
Nightjars (5)							
Orioles (6)							
Owls (7)							
Phainopepla							
Rails (2)						■	■
Raptors (1)							
Robin, American							
Shorebirds (25)	■	■	■	■	■	■	■
Shrikes (2)							
Solitaire, Townsend's							
Starling, European							
Storm-petrels (4)	■						
Swallows (1)							
Tanagers (3)							
Terns (2)	■	■		■		■	■
Thrashers (8)							
Thrushes (2)							
Towhees (2)							
Waxwings (1)							
Woodcock, American							
Woodpeckers (16)							
Wrens (1)							
Pigeon-sized							
Anis (1)							
Auks (6)	■	■	■				
Blackbirds (2)						■	■
Chickenlike Birds (3)							
Ducks (2)		■					■

Lakes, Ponds, Rivers•
Streams, Brooks
Southern Wooded Swamps•
Boreal Forest
Eastern Deciduous Forest
Northern Wooded Swamps
Southeastern Pine Forest•
Western Coniferous Forest
Dry Western Woodlands•
Thickets
Open Country•
Deserts, Sagebrush Plains
Mountain Cliffs, Rocky Slopes•
Alpine Tundra and Meadows
Urban Areas
Residential•

Habitat and Size

- *Habitat Abbreviated*
- *Water Birds*

	Open Ocean	Inshore Waters	Rocky Shores, Coastal Cliffs	Beaches, Coastal Dunes	Mudflats, Tidal Shallows	Salt Marshes	Freshwater Marshes
Pigeon-sized continued							
• Gallinule, Purple						■	■
• Grebes (3)	■	■				■	■
• Gulls (2)	■	■		■			■
• Jaegers (1)	■	■					
Jay, Brown							
• Kingfishers (1)		■				■	■
• Long-legged Waders (1)						■	■
Magpies (1)							
Nightjars (1)							
Nutcracker, Clark's							
Owls (1)							
Pigeons and Doves (4)					■		
Raptors (2)				■			
• Shearwaters (5)	■						
• Shorebirds (2)					■	■	■
• Terns (8)	■	■	■	■	■	■	■
Trogon, Elegant							
Woodpeckers (1)							
Crow-sized							
• Auks (3)	■	■	■				
Chickenlike Birds (10)							
• Coot, American		■				■	■
Crows and Ravens (5)		■	■	■	■	■	■
• Ducks (30)		■	■		■	■	■
• Geese (1)						■	■
• Grebes (2)		■					■
• Gulls (12)	■	■	■	■	■	■	■
• Jaegers and Skuas (4)	■	■	■	■			
• Long-legged Waders (2)						■	■
• Moorhen, Common					■		■
Owls (7)				■		■	■
• Rails (2)					■	■	■

Lakes, Ponds, Rivers•
Streams, Brooks
Southern Wooded Swamps•
Boreal Forest
Eastern Deciduous Forest
Northern Wooded Swamps
Southeastern Pine Forest•
Western Coniferous Forest
Dry Western Woodlands•
Thickets
Open Country•
Deserts, Sagebrush Plains
Mountain Cliffs, Rocky Slopes•
Alpine Tundra and Meadows
Urban Areas
Residential•

Habitat and Size

- Habitat Abbreviated
- Water Birds

	Open Ocean	Inshore Waters	Rocky Shores, Coastal Cliffs	Beaches, Coastal Dunes	Mudflats, Tidal Shallows	Salt Marshes	Freshwater Marshes
Crow-sized continued							
Raptors (25)		■		■		■	■
Roadrunner, Greater							
● Shearwaters (5)	■						
● Shorebirds (9)			■	■	■	■	■
● Skimmer, Black		■			■	■	
● Terns (3)		■	■	■	■	■	■
● Tropicbirds (2)	■						
Woodpeckers (1)							
Goose-sized							
● Anhinga		■				■	■
● Boobies (1)	■	■					
Chickenlike Birds (2)							■
● Cormorants (2)	■	■	■				■
● Ducks (2)		■	■				
● Frigatebird, Magnificent	■	■		■			
● Geese (5)		■	■			■	■
● Gulls (6)	■	■	■	■	■	■	■
● Long-legged Waders (12)		■			■	■	■
● Loons (2)	■	■					
Owls (1)							
Raptors (2)							
Very Large							
● Albatrosses (2)	■						
● Boobies and Gannets (2)	■	■	■				
● Cormorants (3)	■	■	■	■	■		■
● Geese (1)		■				■	■
● Long-legged Waders (6)		■			■	■	■
● Loons (2)	■	■					
● Pelicans (2)		■	■	■			■
Raptors (3)		■					
● Swans (3)		■				■	■
Turkey, Wild							

Lakes, Ponds, Rivers•
Streams, Brooks
Southern Wooded Swamps•
Boreal Forest
Eastern Deciduous Forest
Northern Wooded Swamps
Southeastern Pine Forest•
Western Coniferous Forest
Dry Western Woodlands•
Thickets
Open Country•
Deserts, Sagebrush Plains
Mountain Cliffs, Rocky Slopes•
Alpine Tundra and Meadows
Urban Areas
Residential•

Behavior

Behavior

For easy reference, bird behavior is divided into three general categories: water, land, and flight.

You may wonder why I have picked behavior as the next step in field identification. Why not more conspicuous field mark categories such as shape or color? It is true that at close range both the shapes and the colors of birds are generally more obvious than their behavior, but a bird's behavior can sometimes be seen at distances so great that you can't make out its color and shape, or under lighting conditions in which colors cannot be seen. The split-second identifications made by expert birders are very often based on behavior rather than color or shape. The sooner you begin to pay attention to what birds do, the sooner you too will be able to name birds instantly.

I wish I had known this when I first began to watch birds. I had been birding for many years before I found out that one can distinguish the Greater Yellowlegs and Lesser Yellowlegs, two shorebirds that are almost identical in shape and color, from the window of a speeding train. The Lesser always probes for its food or picks it from the surface, but the Greater often feeds by sweeping its bill back and forth through the water. This distinctive behavior can be seen from so far away that it can be used to spot a Greater Yellowlegs even when shape and color are all but invisible. It was nearly as long before I learned that in the northern United States, a distant egret sprinting through the shallows had to be a Snowy. The time to begin learning about bird behavior and using it as a valuable source of information is when you first begin watching birds.

What Is Bird Behavior?

Bird behavior includes everything birds do, even the most intimate details of their nesting and courtship activities. But it is their more obvious actions, their different methods of swimming, foraging, and flying, that are useful as field marks. In the following pages, I have organized the pictures, text, and charts of this highly visible bird behavior into three general categories: water, land, and flight. The first category includes behavior that you will see in birds on or near water; it includes swimming, diving from the surface, and a variety of feeding activities, both in the water and along its edges. The second category consists of behaviors you will see in birds that are found on the ground or in bushes or trees. In addition to feeding behavior, this category includes striking actions, such as head-bobbing, wing-flicking, and tail-fanning, behavior that is very noticeable in certain birds but that seems to have no direct relation to obtaining food. The third category consists of aerial behavior—soaring, flying in formations or bunches, hawking for insects from a perch, and the many other activities that make some birds easy to identify when they are on the wing. Study the pictures and descriptions so that you will understand the many kinds of behavior included in each of these three categories. Bear in mind that most of the water behavior refers to birds known as water birds, while the land section refers strictly to land birds. Flight encompasses both groups.

Behavior and Identification

As you will learn, many birds have behavior—in the water, on land, or in flight—so distinctive that you can use it to narrow the range of possibilities immediately to a single group or to one species. If I am on

The water category covers bird activities that occur either on or near water. In salt- or freshwater habitats, you can observe birds behaving in the manner illustrated on the next pages. The land category refers to bird actions on the ground, in bushes, or in trees. These behaviors occur in forest, open, urban and residential habitats, and in marshes.

The flight category describes aerial activities everywhere—either over water or land. These behaviors can be seen in all habitats.

a mudflat or beach, I can count on seeing some birds that alternately run and stop, and others that walk or run more or less continuously over the mud or sand; if a bird runs and stops, I know immediately which group it belongs to. At the beach, if I see a small party of sparrow-sized birds racing down the sand after a retreating wave, this tells me at once what species they are. Most of the long-legged waders hold their long neck fully extended when they are flying, but the members of one group fly with the neck folded back and the head resting between the wings; this single flight behavior identifies these birds at once.

On land, a robin-sized bird hitching its way up the trunk of a tree can belong to only one group, since the other birds that frequently cling to tree trunks are either larger, smaller, or can scramble easily over the bark in any direction, often moving headfirst down the trunk. Noisy scratching in the leaves on the forest floor might be a squirrel, but if it is a bird, there are only a few species it can be. A bird darting out from a perch to catch flying insects is almost certain to be a member of one of two groups of songbirds that catch insects in this way.

How to Study Bird Behavior

Nothing could be easier or more enjoyable than beginning to learn about bird behavior. All you need is your notebook, a pencil, and a little patience. Binoculars are helpful, but not necessary. In your backyard, a city park, or in any habitat you visit, take a good look at the first bird you see. After noting its size, ask yourself what it is doing. If the bird is near the shore, is it probing for food or scavenging on a beach? Is it feeding as it swims? Or diving for fish? Or wading along the margin of a pond, searching for a frog to seize in its bill? If it is on the ground in a park, meadow, or forest, is it walking, running, or hopping? Does the bird seem to be foraging? If so, is it finding insects, fruit, or seeds? Is it feeding at flowers? If the bird is in flight, are its wingbeats fast or slow? Is its flight path straight, bounding, or erratic and darting? Is the bird hovering? Take detailed notes; every note you write down about what a bird is doing will increase your familiarity with that species and will help you to identify the bird if you don't yet know what it is. Once you have named the bird, your notes on its behavior will help you recognize it at once next time.

It has justly been said that the more we understand, the more we see, and the more we see, the more we enjoy. Even when you already know what the species is, knowing about bird behavior will enable you to appreciate things most people overlook. When a flock of gulls is resting on a sandbar, the birds almost always face into the wind, so that their plumage is not ruffled by the breeze. This behavior is not very useful in identification, but being aware of it will not only tell you the direction of the wind at a glance, but can add to your enjoyment of the birds themselves, because you will see and understand this simple action they have taken to make themselves more comfortable. A knowledge of bird behavior can turn what would otherwise be an ordinary stroll along a beach or walk through the woods into a rich harvest of fascinating observations.

Behavior on or near Water

Swimming
Many birds swim on the surface of the water. Most of these ride fairly high, like a gull, with much of the body as well as the head, neck, and tail clearly visible. A few birds swim with the body lower.

Tundra Swan

Head-pumping While Swimming
Three water birds—the Common Moorhen, Purple Gallinule, and American Coot—can be identified easily by their habit of pumping their heads back and forth while swimming. Experts are not sure why they do this, but it has been suggested that this head movement helps them keep their surroundings in focus.

Common Moorhen

Sinking
A few groups of water birds, all of them birds that dive, can sink gradually out of sight to avoid danger. By compressing the plumage of the body, they force out trapped air that would otherwise make them more buoyant.

Ruddy Duck

Swimming with Body Low in Water
Some swimming birds ride with the body so low that water washes over the back, and sometimes only the head and neck show above the surface. All of these birds are rather heavy-bodied divers.

Western Grebe

All the behaviors illustrated and
described on this page appear on
the charts under Swimming.

Wood Duck

Western Gull

Common Moorhen

Common Moorhen

Ruddy Duck

Ruddy Duck

Double-crested Cormorant

Red-throated Loon

Behavior on or near Water

Swimming with Only Head Visible

A few diving birds may often be seen with only the head above the surface, holding a newly captured fish. When alarmed, a grebe may hide by resting in the water with only its head visible.

Anhinga

Diving from Surface

Some water birds seek their food by diving from the surface, and pursuing fish or insects underwater or plucking vegetation from the bottom.

Horned Grebe

Feeding at Surface

Many swimming birds forage by picking food from the surface, without diving. Some of these also find food by reaching beneath the surface. The phalaropes spin around on the water, stirring up small aquatic animals, which they then seize in their bills.

Red-necked Phalarope

Tipping Up

Although they do not dive, many dabbling ducks, geese, and swans obtain much of their food by tipping up—tilting the body forward so that the tail points up, while they reach for aquatic plants growing on the bottom of shallow water.

Northern Shoveler

Look for these behaviors on the
charts under Swimming and
Feeding.

Anhinga

Anhinga

Horned Grebe

Horned Grebe

Northern Shoveler

Trumpeter Swan

Blue-winged Teal

Mallard

Behavior on or near Water

Holding Neck in an S-curve
When herons and flamingos stand at the water's edge, or stalk in the shallows, they usually hold their necks in an S-curve, rather than straight up as do most other long-legged waders. Perched or swimming, the Anhinga also holds its slender neck in an S-curve.

Tricolored Heron

Bill-sweeping
A few wading birds feed by rapidly sweeping their bills from side to side, straining small aquatic animals out of the shallow water.

American Avocet

Probing Mud or Sand
Many marsh-dwelling birds and shorebirds use their bills to probe for food, instead of snatching it from the water's surface. The differing lengths of the bills of these birds allow them to forage at different depths, so a number of species can feed on the same shore without competing.

Western Sandpiper

Turning Over Stones on Beach
The turnstones are named for their habit of quickly overturning small stones, shells, and pieces of wood, and catching the sand fleas hiding underneath. But only the Ruddy Turnstone commonly does this; the Black Turnstone lives on rocky shores, not beaches, and seldom hunts for food in this way.

Ruddy Turnstone

These behaviors appear on the
charts under Feeding.

Tricolored Heron

Tricolored Heron

American Avocet

American Avocet

Short-billed Dowitcher

Marbled Godwit

Ruddy Turnstone

Ruddy Turnstone

Behavior on or near Water

Running and Stopping
Plovers have a very distinctive way of hunting for food: They run quickly for several paces and stop, watching the mud or sand for small prey. The instant they see a movement ahead, they dart forward and snatch up the catch. Soon they repeat this process.

Wilson's Plover

Chasing Retreating Waves
Unlike most shorebirds that forage on mudflats, the Sanderling races down the beach after retreating waves, capturing small sand-dwelling creatures before they burrow back out of sight.

Sanderling

Sprint-wading
While most wading birds walk through the water, watching for prey, a few are very active. These birds sprint about so that small animals are frightened out of their hiding places. This "sprint-wading" is so distinctive that you can identify such a bird at a great distance.

Snowy Egret

Wading
In contrast to the birds that swim, many birds have long legs and seek their prey by wading. Larger waders, with longer legs, can forage in deeper water than small, short-legged sandpipers and other shorebirds.

Wood Stork

Look for these behaviors on the
charts under Gait.

Semipalmated Plover

Wilson's Plover

Sanderling

Sanderling

Greater Yellowlegs

Lesser Yellowlegs

Great Blue Heron

Black-necked Stilt

Behavior on or near Water

Diving from Air
A number of fish-eating birds catch their prey by diving into the water from the air. Most of these birds seize fish in their bills, but the Osprey and Bald Eagle capture prey in their talons.

Brown Booby

Skimming
The Black Skimmer probably has the most distinctive feeding method of any water bird: Flying close to the surface, it plows its long lower mandible through the water in pursuit of small fish; the moment the lower mandible strikes a fish, the bill snaps shut.

Black Skimmer

Wing-raising
A number of sandpipers sometimes stand with one or both wings raised, so that the wing linings are visible. The wings may be held up only briefly, just after a bird lands, or it may form part of a courtship display you may see in spring, as the birds travel north to their breeding grounds.

Wilson's Phalarope

Wing-spreading
Cormorants and the Anhinga are diving birds, but their plumage is not water-repellent. To keep their feathers from becoming waterlogged, these birds must perch on rocks or branches, where they may be seen spreading their wings to dry.

Great Cormorant

You will find these behaviors on
the charts under Wings.

Brown Pelican

Red-billed Tropicbird

Black Skimmer

Black Skimmer

Hudsonian Godwit

Dunlin

Double-crested Cormorant

Anhinga

Behavior on Ground, in Bushes, and in Trees

Flushing
Many ground-dwelling birds rely on their streaked or brown plumage to conceal them when a predator approaches, and then flush, bursting into the air at the last moment and leaving the startled predator far behind.

Western Meadowlark

Wing-flashing
The Northern Mockingbird exhibits a distinctive behavior known as wing-flashing. The bird stands on the ground and slowly spreads its wings outward and forward in a series of quick, jerky motions. It then folds its wings, runs forward, and repeats the process. Wing-flashing probably flushes insects from hiding.

Northern Mockingbird

Wing-flicking
Several birds flick their wings as they forage in trees and bushes. Some do this only when nervous, while others flick their wings more or less continuously.

Golden-crowned Kinglet

Head-bobbing
A few birds frequently bob their heads up and down, especially when they are curious or slightly alarmed. This behavior probably enables them to gauge the distance of objects around them.

Burrowing Owl

These behaviors appear on the charts under Wings and Gait.

Short-eared Owl

Wild Turkey

Northern Mockingbird

Northern Mockingbird

Golden-crowned Kinglet

Golden-crowned Kinglet

Burrowing Owl

Burrowing Owl

Behavior on Ground, in Bushes, and in Trees

Walking or Running
Most songbirds adapted to life in trees hop, even when they come to the ground, but ground-dwelling species generally walk or run. Be careful with the sparrow-sized thrushes, however; these birds hop over the ground so fast that it often looks as though they are running, but only the larger thrushes actually walk or run.

Greater Roadrunner

Running and Stopping
Many ground-dwelling birds walk or run, but the American Robin, Northern Mockingbird, and Varied Thrush have a distinctive way of foraging. They run forward for several paces, then stop and scan the ground for insects or other prey. If they see something to eat, they quickly seize it; if not, they move forward and try again.

Northern Mockingbird

Noisily Scratching in Leaves
Birds often scratch in dead leaves looking for insects and seeds, but a few scratch so noisily that you will notice them right away, even before you actually see them. Most of these birds scratch with their feet, but thrashers use their bills.

Rufous-sided Towhee

Flipping Over Leaves
A few ground-dwelling birds forage by moving over the forest floor or along the margins of streams, quickly flipping leaves over with their bills and seizing insects hiding underneath.

Wood Thrush

Look for these behaviors on the
charts under Gait and Feeding.

Western Meadowlark

White-throated Sparrow

Varied Thrush

American Robin

Rufous-sided Towhee

Rufous-sided Towhee

Wood Thrush

Wood Thrush

Behavior on Ground, in Bushes, and in Trees

Feeding at Flowers
Hummingbirds are well known for visiting flowers to obtain nectar and small insects. Other birds also visit flowers for nectar, and some, mainly the Cedar Waxwing, finches, and the House Sparrow, can be seen at flowers, busily eating the petals.

Hooded Oriole

Probing Pinecones
A few songbirds regularly probe pinecones for seeds or hibernating insects.

Red Crossbill

Hitching up Tree Trunks
Most tree-clinging birds have stiff tail feathers that they use as props while they grip the bark of a tree with their feet. These birds usually hitch their way up a tree trunk, and never position themselves head-downward.

Ladder-backed Woodpecker

Climbing down Tree Trunks
Unlike most tree-clinging birds, nuthatches have normal tail feathers and very strong feet that seemingly enable them to defy gravity by creeping in any direction—whether upward or downward—on a tree trunk. The Black-and-white Warbler also climbs up and down trees.

Red-breasted Nuthatch

You will find these behaviors on
the charts under Feeding.

Cedar Waxwing	Ruby-throated Hummingbird

White-winged Crossbill	Common Redpoll

Red-headed Woodpecker	Hairy Woodpecker

White-breasted Nuthatch	Pygmy Nuthatch

Behavior on Ground, in Bushes, and in Trees

Slow Foraging in Foliage
Since most small songbirds forage in foliage very rapidly, it is often easy to pick out vireos, and those few warblers that forage more slowly, taking time to peer under leaves for insects, and moving deliberately from branch to branch.

Warbling Vireo

Rapid Foraging in Foliage
Many small songbirds move very rapidly through the foliage of trees in search of insects. This behavior is usually associated with warblers, but even nuthatches, when not creeping about on trunks and branches, forage in this way.

Black-throated Green Warbler

Skulking
Some birds are characteristically shy and likely to try to stay out of sight by skulking in dense vegetation. Many of these birds can be lured into view if you make a squeaking noise, like a bird in distress.

Yellow-rumped Warbler

Cocking Tail Over Back
Several birds that live in thickets or trees often carry their tails held either straight up or tilted forward over the back.

Marsh Wren

These behaviors appear on the charts under Feeding, Skulking, and Tail.

Bell's Vireo

Bay-breasted Warbler

Black-and-white Warbler

Prothonotary Warbler

Song Sparrow

Henslow's Sparrow

Marsh Wren

Cactus Wren

Behavior on Ground, in Bushes, and in Trees

Tail-fanning
A small number of birds habitually fan their tails as they forage in foliage. Most of these birds have patches of white, yellow, or orange in the tail, making this behavior quite noticeable.

Magnolia Warbler

Tail-bobbing
As they walk over the ground, a number of birds bob their tails up and down more or less rhythmically and continuously. Why they do this is not clear, but the habit is an excellent field mark.

Kirtland's Warbler

Tail-flicking
Phoebes are well known for flicking their tails, but other birds also have this trait. Some do it only when nervous, others even when foraging quietly.

Black Phoebe

Pointing Tail Downward
Although almost any songbird may perch for a moment with its tail pointed downward, some birds hold their tails in this position so frequently that the habit can be helpful in identification.

Sulphur-bellied Flycatcher

ook for these behaviors on the
arts under Tail.

American Redstart

Bewick's Wren

venbird

Water Pipit

ilson's Warbler

Hermit Thrush

east Flycatcher

Vermilion Flycatcher

Flight Behavior

Soaring
To save energy during long migrations, or while circling high in the air and searching for food, many birds soar on outstretched, motionless wings. When air currents are deflected upward by long mountain ridges, some of these birds can travel for miles without a single wingbeat.

Bald Eagle

Dihedral Soaring
While most soaring birds hold their outstretched wings flat, a few raptors soar with the wings held in a shallow angle or V called a dihedral—a term borrowed from aviation. As these birds soar on dihedral wings, they usually appear unsteady, swaying from side to side.

Turkey Vulture

Flapping and Coasting
Several birds alternate periods of flapping with long glides or coasting. The water birds use this method to conserve energy while traveling long distances. Birds of prey often soar, but interrupt their soaring with frequent wingbeats to maintain their altitude.

American White Pelican

Swallowlike Flight
The swallows are not the only birds whose flight is light and buoyant, with frequent skimming and swooping; terns and storm-petrels can also be recognized by their swallowlike flight. Swifts, which superficially resemble swallows, have a very different and distinctive flight—very fast on rapidly beating wings.

Caspian Tern

The behaviors illustrated here and on the next pages appear on the charts under Flight Style.

Bald Eagle

Bald Eagle

Turkey Vulture

Turkey Vulture

American White Pelican

American White Pelican

Forster's Tern

Common Tern

Flight Behavior

Beeline Flight
Some birds have a regular, rhythmic, almost mechanical wingbeat and a perfectly straight flight path, as if the bird were being pulled rapidly along a taut wire. This "beeline" flight is characteristic of some large birds with relatively slow wingbeats, as well as some of the tiny hummingbirds.

White-winged Dove

Flying with Long Neck Extended
Nearly all birds with long necks fly with the neck fully extended, not folded back or with a kink. This field mark is especially useful in sorting out the long-legged waders.

Roseate Spoonbill

Flying with Neck Folded
Unlike most long-necked birds, the herons, including the egrets and bitterns, fly with the neck folded back so that the head rests between the wings. Pelicans also fold their necks back, perhaps to help support the weight of their huge bills.

Brown Pelican

Flying with Kink in Neck
A few long-necked birds fly with a distinct kink or crook in the neck. This feature distinguishes the Double-crested Cormorant from all other cormorants, and is also noted in the Anhinga.

Anhinga

Mourning Dove

Common Loon

Sandhill Crane

White Ibis

Great Egret

Tricolored Heron

Double-crested Cormorant

Double-crested Cormorant

Flight Behavior

Flying with Rapid Wingbeats
Smaller birds tend to have more rapid wingbeats than larger ones, but some birds have a fast wingbeat because their wings are very small or short in relation to the size of the body. Although swifts have long wings, the bend of the wing is very close to the body, and they also beat their wings rapidly.

Common Eider

Flying with Slow Wingbeats
The largest birds beat their wings very slowly. This is a valuable clue in judging the size of a bird at a great distance.

Great Egret

Stiff-winged Flight
When some seabirds flap their wings in normal flight, the angle of the wing does not bend, creating a striking, stiff-winged effect. The birds use their stiffly held wings to coast on small air currents over the waves.

Laysan Albatross

Loose-winged Flight
In contrast to birds that fly with stiff, unbending wings, some birds seem to have very loose joints, and fly with gangling, almost floppy wingbeats.

Short-eared Owl

Wild Turkey

Surf Scoter

Great Egret

Great Egret

Laysan Albatross

Laysan Albatross

Short-eared Owl

Short-eared Owl

Flight Behavior

Flying in Bunches
When flying long distances, a few large birds, and nearly all small birds that move in flocks, tend to travel in loose or compact bunches, and not in formations, such as V's, arcs, or waves.

Dunlin

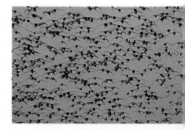

Erratic Flight
The erratic, darting or twisting flight of some species makes them easy to distinguish from similar birds that have a straight flight path.

Blue-winged Teal

Flying in Formation: V, Arc, or Wave
During prolonged flight, flocks of many species tend to fly in a pointed V, or in a rounded arc or wave. Although no species invariably flies in a particular formation, flying in any formation distinguishes these birds from those that habitually fly in bunches.

Canada Goose

Flying in Formation: Line
Flocks of birds are commonly seen flying in a line. Many of these birds also fly in a V, arc, or wave. Flocks on their way to a roost often travel in a line-formation.

Glossy Ibis

These behaviors appear on the
charts under Flight Style and
Group Flight.

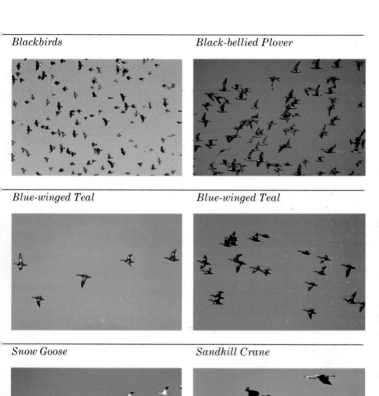

Blackbirds

Black-bellied Plover

Blue-winged Teal

Blue-winged Teal

Snow Goose

Sandhill Crane

Sandhill Crane

Canada Goose

Flight Behavior

Follow-the-leader Flapping
Some water birds that fly in lines may coast for several seconds without flapping their wings. When the lead bird flaps its wings, each bird in the line then flaps its wings in turn, as if copying the bird just in front of it. A wave of flapping moves down the line.

American White Pelican

Simultaneous Banking
In some species, the birds in a fast-moving flock bank in a highly coordinated way. All the birds change position at almost exactly the same instant. A distant flock of sandpipers seems to appear and disappear, as the birds reveal their dark underparts then their pale undersides to the observer.

Green-winged Teal

Mobbing Predators
Many birds scold or attack hawks, owls, and other predators, even those larger than themselves. Birds "mob" predators not only to protect their nests, but to keep the predator under surveillance, and perhaps to teach younger birds to recognize dangerous birds and mammals.

Western Gull mobbing Osprey

Robbing Food from Other Birds
Some birds are aerial pirates, robbing other birds of food rather than capturing prey of their own. The usual victims are gulls and terns, but the Bald Eagle generally takes fish captured by the Osprey. Gulls often rob one another of food, sometimes attacking birds of their own species.

Long-tailed Jaeger

Look for these behaviors on the
charts under Group Flight and
Feeding.

American White Pelican

American White Pelican

Green-winged Teal

American Crow mobbing barn-owl

Scissortail mobbing hawk

Long-tailed Jaeger

Long-tailed Jaeger

Flight Behavior

Swooping Up to a Perch
While most predatory birds are liable to approach a perch flying close to the ground, then rise suddenly to alight, a few species do this so often that it is a good field mark.

Northern Hawk-Owl

Whistling Wings
Several birds produce a loud whistling or trilling sound with their wings as they fly, often enabling a skilled birder to name the species without even having seen it.

American Woodcock

Hawking for Insects from a Perch
Flying insects are an important source of food for many birds. One method of catching them is to dart out from a perch and pluck them out of the air. Almost any bird will do this at times, but some "fly-catch" so often that it is a good clue in identification.

Great Crested Flycatcher

Hovering
Hummingbirds are famous for hovering, poised in front of flowers on almost invisible wings. But several other birds beat their wings rapidly to maintain a stationary position in the air, as they scan the water or ground below for prey, or peer at foliage in search of insects.

Costa's Hummingbird

These behaviors appear on the
charts under Feeding or the first
two categories given below.

Northern Hawk-Owl

Northern Hawk-Owl

Barrow's Goldeneye

Broad-tailed Hummingbird

Great Crested Flycatcher

Great Crested Flycatcher

Rufous Hummingbird

Broad-billed Hummingbird

How to Use the Behavior Charts

The ten behavior charts are divided into two categories, depending on whether the behavior took place in or near water, or on or over land.

Now that you have studied the previous pages, you know it is not only important to note that a bird is swimming, but also how it swims; not merely that it is on the ground, but whether it is walking, running, or hopping; and not just that it is flying, but whether it is hovering, soaring, or flying with rapid wingbeats. You are now ready to take your notebook and go birding.

Tips for the Field

I suggest that you visit a habitat near your home, one that you visit frequently. Find a comfortable place to sit in an inconspicuous spot and avoid making sudden movements. You can do this at any hour, but land birds and birds in marshes are least active in the middle of the day, so it is best to go out early in the morning or late in the afternoon. Most water birds tend to be active all day, so it doesn't matter what time you visit their habitats. The birds and other wildlife will become accustomed to your presence. Soon you will see one of the most interesting things the outdoors has to offer: birds behaving naturally in their natural surroundings.

You have already noted the habitat you are in; now you should record the size of the first bird that comes along. Then, quietly writing in your notebook, describe what the bird is doing.

Using the Charts

To help you match behavior with bird groups or species, I have devised ten charts on the following pages. On the charts, the various behaviors are listed in the column on the left; the bird groups and species that display these behaviors are listed at the top, with similar behaviors next to each other. Near the bottom of the page, you will also find the major habitat groups and the size category. By using behavior, many groups can be broken down into individual species, because a great many species have distinctive behaviors not shared by all members of their group.

Arrangement of Behavior Field Marks

In the illustration section, we divided bird behaviors into three general categories, depending on whether the behavior took place in or near water, on land, or in flight. This same organization is reflected in the charts. But since flight can be observed on or near water as well as over the ground, I have divided the charts between the categories water and land, and listed flight behavior on all charts, noting the beginning of the flight section with an arrow.

Five behaviors appear on the chart but are not illustrated in the preceding sections: drumming; bounding flight; mothlike flight; hanging upside down; and walking on lily pads. Drumming is simply loud tapping on wood; woodpeckers most often engage in this behavior. Unlike birds that fly in a straight line, those with bounding flight seem to bounce through the air; woodpeckers are also known for this flight style. Nightjars are noted for their mothlike flight. Chickadees and titmice hang upside down from twigs while foraging. Walking on lily pads is seen in only a very few marsh birds.

Using Behavior, Size, and Habitat

In the first four charts, you will find birds that are likely to be seen in

On or near Water	On or over Land
Sparrow-sized and Robin-sized	Very Small
Pigeon-sized	Sparrow-sized, three charts
Crow-sized	Robin-sized
Goose-sized and Very Large	Pigeon-sized, Crow-sized,
	Goose-sized, and Very Large

and around water. The birds are organized by size, beginning with sparrow- and robin-sized birds and ending with very large birds. Most of these birds are water birds; however, a black dot before the name indicates that this bird or group is considered a land bird. The next six charts show ground behavior, and the birds covered here are also organized according to size. Because sparrow-sized birds are so numerous, the birds in this yardstick category have been further divided, according to how easy it is to find them, and whether they are most often found feeding on the ground, or in bushes, trees, or in the air. Here a black dot next to the name indicates a water bird. Most charts contain more than one size category, so be sure to check the size information at the bottom of the page, as well as the habitats.

Giveaway Behavior

While some behaviors are found in many birds, and you may need other field marks to sort out the species, certain behaviors are so distinctive and performed by so few birds that you can immediately pinpoint groups or species the moment you see one of these behaviors. These "giveaway" behaviors are indicated in the charts by a red square. You should be on the alert for giveaway behaviors, because they will often lead you quickly to an identification. But do not neglect the other behaviors shown in the charts; in combination with other field marks, like shape or color, they can be equally useful.

Once you have been in the field and have made notes on the habitat, size, and behavior of birds, you are ready to use your field guide to verify your identification. As you'll see in the companion field guides, habitat and size are noted on the top of each picture-and-text account and the giveaway behaviors are mentioned in the first paragraph of the Field Marks section.

How the System Works

You may have been at the beach and noted that you watched a robin-sized bird. Checking the habitat and size charts at home, you discover that only two groups of robin-sized birds are likely to be found at a beach—a large number of shorebirds and a few terns. You decide the bird was likely to have been a shorebird, and, according to your notebook, it was walking along, flipping over stones and shells. Referring to the behavior chart, you see that only one bird, the Ruddy Turnstone, behaves this way. A quick glance at the illustration in your field guide confirms your identification.

Let's assume you saw another shorebird, also robin-sized, that ran along for several paces and then stopped; in a few seconds, it ran forward again, and this time picked up something from the surface of the sand. The behavior chart tells you that this bird must be one of the plovers. Now refer to your field guide; you will probably be able to decide which plover you have seen.

What if you saw a larger, crow-sized shorebird probing the sand? The behavior chart lists several crow-sized birds that probe sand. It seems that you are far from identifying this bird, but then you notice that it has a long, strongly downcurved bill. Move on to Shape and Posture, our next chapter. With a striking field mark like the downcurved bill, you will be able to name the bird very quickly.

Behavior 1
On or near Water

Sparrow- and Robin-sized
■ *Giveaway Field Mark*
► *Flight*
● *Land Birds*

1. Little Gull
2. Storm-petrels
3. Terns
● 4. Swallows
5. Rails
6. Plovers
7. Ruddy Turnstone
8. Northern Jacana

	1	2	3	4	5	6	7	8
Swimming	■							
Sinking								
Body low in water								
Only head visible								
Diving from surface								
Feeding: At surface	■	■						■
Probing mud or sand					■		■	■
Turning over stones						■		
Gait: Walking or running					■	■	■	■
Running and stopping						■		
Chasing retreating waves								
Sprint-wading								
Wading					■			■
Walking on lily pads							■	
Head-bobbing								
Tail: Bobbing								
Wings: Diving from air			■					
Raising								■
Flushing					■			
► **Flight Style:** Swallowlike		■	■	■				
Rapid wingbeats					■	■	■	
Stiff-winged								
Erratic				■				
Group Flight: In bunches						■	■	
Formations						■	■	
Simultaneous banking							■	
Feeding: Mobbing predators			■					
Hovering		■	■					
Whistling Wings								
Saltwater Habitats	■	■	■	■	■	■	■	
Freshwater Habitats	■			■	■	■		■
Sparrow-sized		■		■	■	■		
Robin-sized	■	■	■		■	■	■	■

9. Sandpipers
10. Sanderling
11. Common Snipe
12. American Woodcock
13. Spotted Sandpiper
14. Solitary Sandpiper
15. Lesser Yellowlegs
16. Phalaropes
17. Wilson's Phalarope

● 18. American Dipper
19. Auks
20. Least Grebe
21. Green Kingfisher
● 22. Waterthrushes

9	10	11	12	13	14	15	16	17	18	19	20	21	22
							■	■	■	■	■		
											■		
											▨		
											■		
									■	■	■		
							■	■	■				
■	■	■	■	■	■	■	■	■	■				
■	■	■	■	■	■	■	■	■	■			■	
	▨												
						▨		▨					
■	■	■		■	■	■	■	■	■				
				■	■								
				▨	▨				▨				▨
											▨		
■								■					
		■	■										
					■								
■	■	■	■		■	■	■	■		■	■	■	
				▨									
■	■	▨		■	■		■	■					
■	■							■					
■	■							■					
■	■						■	■					
											■		
			▨										
▨	▨			▨	▨	▨	▨	▨		▨			
▨		▨	▨	▨	▨	▨	▨	▨	▨		▨	▨	▨
■				■					■				■
■	■	■	■		■	■	■	■		■	▨	■	

Behavior 2
On or near Water

Pigeon-sized
■ *Giveaway Field Mark*
▶ *Flight*
● *Land Birds*

1. Auks
2. Grebes
3. Bufflehead
4. Masked Duck
5. Shearwaters
6. Gulls
7. Long-tailed Jaeger
8. Terns

	1	2	3	4	5	6	7	8
Swimming	■	■	■	■	■	■		
Head-pumping								
Sinking		■						
Body low in water		■		■				
Only head visible		■						
Diving from surface	■	■	■	■				
Feeding: At surface					■	■		
Bill-sweeping								
Probing mud or sand								
Gait: Walking or running								
Sprint-wading								
Wading								
Walking on lily pads								
Head-bobbing								
Skulking								
Wings: Diving from air								■
▶ **Flight Style:** Soaring								
Flapping and coasting					■			
Swallowlike							■	■
Rapid wingbeats	■	■	■					
Slow wingbeats								
Stiff-winged					■			
Group Flight: In bunches				■	■			
Formations		■						
Feeding: Mobbing predators						■		■
Robbing food from other birds						■	■	
Hovering								■
Saltwater Habitats	■	■	■		■	■	■	■
Freshwater Habitats		■	■	■		■		■
Pigeon-sized	■	■	■	■	■	■	■	■

9. Purple Gallinule
10. Greater Yellowlegs
11. Least Bittern
12. Belted Kingfisher
● 13. Sharp-shinned Hawk
● 14. Merlin

9	10	11	12	13	14
■					
■					
■					
	■				
■	■				
■	■	■			
	■				
	■	■			
■					
	■				
		■			
			■		
			■	■	
			■		
	■				■
		■			
			■		
■	■	■	■	■	■
■	■	■	■		
■	■	■	■	■	■

Behavior 3
On or near Water

Crow-sized
- ■ *Giveaway Field Mark*
- ► *Flight*
- ● *Land Birds*

1. Black Skimmer
2. Tropicbirds
3. Terns
4. Jaegers and Skuas
5. Shearwaters
6. Gulls
7. Black-legged Kittiwake
8. Auks

	1	2	3	4	5	6	7	8
Swimming					■	■	■	■
Head-pumping								
Sinking/Body low in water								
Diving from surface								■
Feeding: At surface					■	■		
Tipping up								
Neck in S-curve								
Bill-sweeping								
Probing mud or sand								
Gait: Walking or running								
Wading								
Wings: Diving from air		■	■				■	
Skimming	■							
Raising								
Flushing								
► **Flight Style:** Soaring						■	■	
Flapping, coasting/Stiff-winged					■			
Swallowlike			■					
Beeline								
With long neck extended								
With neck folded								
Rapid wingbeats		■						■
Erratic		■		■				
Group Flight: In bunches					■			
Formations								
Simultaneous banking								
Feeding: Mobbing predators			■			■	■	
Robbing food from other birds				■		■		
Hovering			■					
Whistling Wings								
Saltwater Habitats	■	■	■	■	■	■	■	■
Freshwater Habitats			■			■		
Crow-sized	■	■	■	■	■	■	■	■

9. Grebes
10. Ruddy Duck
11. Sea and Bay Ducks
12. Harlequin Duck
13. Goldeneyes
14. Mergansers
15. Dabbling Ducks
16. Teals
17. Whistling-ducks
18. Ross' Goose
19. American Coot
20. Common Moorhen
21. Rails
22. Oystercatchers and
    Black-necked Stilt
23. American Avocet
24. Sandpipers
25. Herons
● 26. Osprey

	9	10	11	12	13	14	15	16	17	18	19	20	21	22	23	24	25	26
	■	■	■	■	■	■	■	■	■	■	■	■						
											▦	▦						
	▦	▦																
	■	■	■	■	■	■					■							
							■	■	■	■	■	■						
							▦	■	■	▦								
																	▦	
															▦			
											■	■	■			■		
								■		■	■	■	■	■	■	■		
											■	■	■	■	■	■		
																		■
																■		
													■					
																		■
					▦													
	■							■	■									
																	▦	
	■	■	■	■	■	■					■	■	■					
				■				■										
			■	■	■	■	■	■							■		■	
		■	■	■	■	■			■								■	
							■											
																		■
				▦														
	▦	▦	▦	▦	▦	▦	▦	▦	▦	▦	▦	▦	▦	▦	▦	▦	▦	▦
	▦	▦	▦	▦	▦	▦	▦	▦	▦	▦	▦	▦	▦	▦	▦	▦	▦	▦
	■	■	■	■	■	■	■	■	■	▦	■	■	■	■	■	■	■	■

Behavior 4
On or near Water

Goose-sized and Very Large
- *Giveaway Field Mark*
► *Flight*
● *Land Birds*

1. Boobies and Northern Gannet
2. Magnificent Frigatebird
3. Albatrosses
4. Gulls
5. Loons
6. Anhinga
7. Cormorants
8. Double-crested Cormorant

	1	2	3	4	5	6	7	8
Swimming			■	■	■	■	■	■
Sinking					■	■		
Body low in water/Only head visible					▨	■	▨	■
Diving from surface					■	■	■	■
Feeding: At surface			■	■				
Tipping up								
Neck in S-curve						■		
Bill-sweeping								
Probing mud or sand								
Gait: Walking or running								
Sprint-wading								
Wading								
Wings: Diving from air	■							
Spreading						▨	▨	■
► **Flight Style:** Soaring		■	■	■		■		
Flapping and coasting				■			■	■
Beeline	■				■			
With long neck extended						■	■	
With neck folded								
With kink in neck						■		▨
Rapid wingbeats								
Slow wingbeats	■		■			■		
Stiff-winged				▨				
Loose-winged								
Group Flight: In bunches								
Formations							■	■
Follow-the-leader flapping							▨	■
Feeding: Mobbing predators				■				
Robbing food from other birds		■		■				
Saltwater Habitats	▨	▨	▨	▨	▨	▨	▨	▨
Freshwater Habitats	▨				▨	▨	▨	▨
Goose-sized	■	■		■	■	■	■	
Very Large	■		■		■		■	■

9. Common Eider
10. Common Merganser
11. Geese
12. Brant
13. Swans
14. American White Pelican
15. Brown Pelican
16. Herons
17. Snowy and Reddish egrets

18. American Bittern
19. Limpkin
20. Ibises
21. Roseate Spoonbill
22. Greater Flamingo
23. Cranes
24. Wood Stork
● 25. Bald Eagle

9	10	11	12	13	14	15	16	17	18	19	20	21	22	23	24	25

Behavior 5
On or over Land

Very Small
■ *Giveaway Field Mark*
► *Flight*

1. Lesser Goldfinch
2. Hummingbirds
3. Broad-tailed Hummingbird
4. Lucy's Warbler
5. Gnatcatchers
6. Kinglets
7. Northern Parula
8. Verdin

	1	2	3	4	5	6	7	8
Wings: Flushing								
Flicking						■		
Gait: Head-bobbing								
Feeding: At flowers	■	■	■					
Probing pinecones								
Climbing down tree trunks								
Hanging upside down						■	■	■
Rapid foraging				■	■	■	■	■
Skulking								
Tail: Cocking over back					■			
Flicking				■	■	■		
► **Flight Style:** Beeline		■	■					
Rapid wingbeats		■	■					
Bounding	■							
Group Flight: In bunches	■							
Feeding: Hovering		■	■		■	■		
Whistling Wings			■					
Freshwater Habitats								
Forest Habitats	■	■	■	■	■	■	■	■
Open Habitats	■	■			■			
Urban and Residential Habitats	■	■			■	■		
Very Small	■	■	■	■	■	■	■	■

9. Nuthatches
10. Bushtit
11. Black-capped Vireo
12. Northern Beardless-Tyrannulet
13. Sedge Wren
14. Winter Wren

Behavior 6
On or over Land

Sparrow-sized
Ground Foragers
■ *Giveaway Field Mark*
▶ *Flight*

1. Sparrows
2. Olive Sparrow
3. Fox Sparrow
4. Green-tailed Towhee
5. Sage Sparrow
6. Connecticut Warbler
7. Dark-eyed Junco
8. Yellow-eyed Junco

	1	2	3	4	5	6	7	8
Wings: Flushing	■	■	■	■			■	■
Flicking								
Gait: Walking or running					■	■		■
Feeding: Noisily scratching in leaves		■	■	◪				
Flipping over leaves								
Slow foraging								
Rapid foraging								
Skulking	■	■	■	■	■	■		
Tail: Cocking over back					■			
Bobbing								
Flicking								
▶ **Flight Style:** Beeline								
Rapid wingbeats								
Bounding								
Group Flight: In bunches								
Simultaneous banking								
Saltwater Habitats	■							
Freshwater Habitats	■							
Forest Habitats	■	■	■	■	■	■	■	■
Open Habitats	■		■	■	■		■	
Urban and Residential Habitats	■		■				■	
Sparrow-sized	■	■	■	■	■	■	■	■

9. Eurasian Skylark
10. Horned Lark
11. Longspurs
12. Snow Bunting
13. Lark Bunting
14. Sprague's Pipit
15. Water Pipit
16. Brown-headed Cowbird
17. Palm Warbler
18. Ovenbird
19. Worm-eating Warbler
20. Kentucky Warbler
21. Doves
22. Thrushes
23. Hermit Thrush

9	10	11	12	13	14	15	16	17	18	19	20	21	22	23

Behavior 7
On or over Land

Sparrow-sized
Easy-to-see Foragers in Bushes,
Trees, or in the Air
- *Giveaway Field Mark*
► *Flight*

1. Swifts
2. Swallows
3. Phoebes
4. Wood-pewees
5. Bluebirds
6. Mountain Bluebird
7. Hummingbirds
8. Orchard Oriole

	1	2	3	4	5	6	7	8
Feeding: At flowers							■	■
Probing pinecones								
Hitching up tree trunks								
Climbing down tree trunks								
Hanging upside down								
Slow foraging								■
Rapid foraging								
Drumming								
Tail: Flicking			■					
Pointing downward			■	■	■	■		
► **Flight Style:** Swallowlike		■						
Beeline							■	
Rapid wingbeats	■						■	
Erratic	■	■						
Bounding								
Group Flight: In bunches								
Feeding: Hawking insects			■	■				
Hovering						■	■	
Saltwater Habitats	■	■						
Freshwater Habitats			■	■				
Forest Habitats			■	■	■	■	■	■
Open Habitats	■	■	■	■	■	■		■
Urban and Residential Habitats	■	■	■					■
Sparrow-sized	■	■	■	■	■	■	■	■

9. House Sparrow	18. Woodpeckers
10. Cedar Waxwing	19. Brown Creeper
11. Finches	20. White-breasted Nuthatch
12. Cassin's Finch	21. Black-and-white Warbler
13. Purple Finch	22. Chickadees
14. House Finch	23. Titmice
15. Pine Siskin	
16. Redpolls	
17. Crossbills	

Behavior 8
On or over Land

Sparrow-sized
Hard-to-see Foragers in Bushes
and in Trees
■ *Giveaway Field Mark*
▶ *Flight*

1. Warblers
2. Kirtland's Warbler
3. Magnolia Warbler
4. Virginia's Warbler
5. Wilson's Warbler
6. Redstarts
7. Hooded Warbler
8. Red-faced and Canada warblers

	1	2	3	4	5	6	7	8
Wings: Flicking					■			
Gait: Walking or running								
Feeding: Hanging upside down								
Slow foraging								
Rapid foraging	■	■	■		■	■	■	■
Skulking								
Tail: Cocking over back								
Fanning			■			■	■	
Bobbing		■						
Flicking				■	■			
Pointing downward								
▶ **Feeding:** Hawking insects					■	■	■	■
Saltwater Habitats	■							
Freshwater Habitats	■						■	
Forest Habitats	■	■	■	■	■	■	■	■
Open Habitats	■				■			
Urban and Residential Habitats	■							
Sparrow-sized	■	■	■	■	■	■	■	■

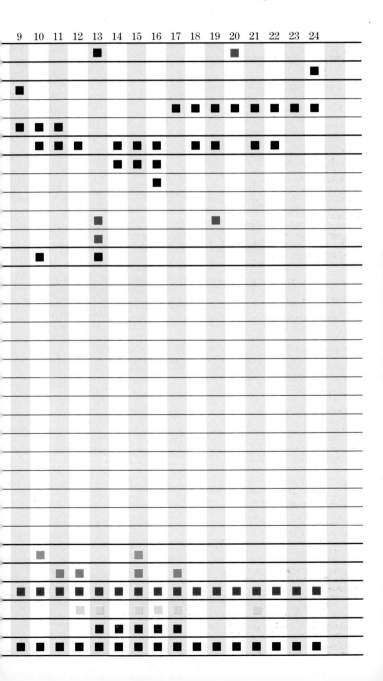

Behavior 9
On or over Land

Robin-sized
■ *Giveaway Field Mark*
▶ *Flight*
● *Water Birds*

1. Townsend's Solitaire
2. Flycatchers
3. Kingbirds and Scissor-tailed Flycatcher
4. Bohemian Waxwing
5. Pine and Evening grosbeaks
6. Jays
7. Pinyon Jay
8. Woodpeckers

	1	2	3	4	5	6	7	8
Wings: Flushing								
Flashing								
Gait: Head-bobbing								
Walking or running							■	
Running and stopping								
Feeding: Noisily scratching in leaves								
Flipping over leaves								
At flowers								
Hitching up tree trunks								■
Slow foraging								
Skulking								
Drumming								■
Tail: Pointing downward	■	■						
▶ **Flight Style:** Swallowlike								
Beeline								
Rapid wingbeats								
Loose-winged								
Erratic								
Bounding				■	■	■		■
Mothlike								
Group Flight: In bunches				■	■			
Simultaneous banking								
Feeding: Mobbing predators			■			■	■	
Hawking insects	■	■	■	■				
Hovering								
Swooping Up to a Perch								
Whistling Wings								
Saltwater Habitats								
Freshwater Habitats			■					
Forest Habitats	■	■	■	■	■	■	■	■
Open Habitats		■	■		■			■
Urban and Residential Habitats	■	■	■		■	■		■
Robin-sized	■	■	■	■	■	■	■	■

9. Thrushes
10. American Robin
11. Burrowing Owl
12. Quails and
   • American Woodcock
13. Thrashers
14. Northern Mockingbird
15. European Starling
16. Blackbirds and Common Grackle
17. Orioles

18. Meadowlarks
19. Towhees
20. Nightjars
21. Nighthawks
22. Purple Martin
23. Cuckoos
24. Doves
25. American Kestrel
26. Shrikes

Behavior 10
On or over Land

Pigeon-, Crow-, and Goose-sized,
and Very Large
■ *Giveaway Field Mark*
▶ *Flight*
● *Water Birds*

1. Pigeons and Rock Dove
● 2. Upland Sandpiper
3. Woodpeckers
4. Northern Hawk-Owl
5. Short-eared Owl
6. Anis
7. Greater Roadrunner
8. Chuck-will's-widow

	1	2	3	4	5	6	7	8
Wings: Raising		■						
Flushing					■			■
Gait: Walking or running	■	■					■	
Feeding: Hitching up tree trunks			■					
Drumming			■					
Tail: Cocking over back							■	
Bobbing				■				
▶ **Flight Style:** Soaring								
Dihedral soaring								
Flapping and coasting						■		
Beeline	■							
Rapid wingbeats	■							
Slow wingbeats					■			
Loose-winged					■	■		
Bounding			■					
Mothlike								■
Group Flight: In bunches	■							
Simultaneous banking	■							
Feeding: Mobbing predators								
Hovering					■			
Swooping Up to a Perch				■				
Saltwater Habitats	■				■			
Freshwater Habitats	■				■			
Forest Habitats	■		■	■		■	■	■
Open Habitats	■	■	■	■	■	■	■	
Urban and Residential Habitats	■		■					
Pigeon-sized	■	■	■	■		■		■
Crow-sized			■		■		■	
Goose-sized								
Very Large								

9. Chickenlike Birds	18. Northern Grosbeak
10. Wild Turkey	19. Black Vulture
11. Clark's Nutcracker	20. Falcons
12. Magpies	21. Golden Eagle
13. Crows	22. California Condor
14. Ravens	23. Northern Harrier
15. Raptors	24. Turkey Vulture
16. Sharp-shinned Hawk	25. Ferruginous and
17. Cooper's Hawk	Rough-legged hawks
	26. Zone-tailed and
	Swainson's hawks

9 10 11 12 13 14 15 16 17 18 19 20 21 22 23 24 25 26

Shape and Posture

Shape and Posture

After you have mastered the habitats, sizes, and behaviors of birds, the next category of field marks to consider is shape and posture. Like behavior, the shape and posture of a bird is often visible at great distances or under conditions in which color and pattern cannot be seen. I find that at close range or in favorable light we all tend to focus on the colors of birds, and for this reason we are likely to overlook the equally useful field marks provided by differences in shape and posture. To avoid being distracted by color, I would like you to consider shape and posture first.

In the chapter on habitat and size, I suggested that you try to ignore shape and posture. This was because we were concentrating on size, and shape and posture can influence our perception of a bird's size, making it more difficult to judge than it need be. Now that you have learned to assign a bird to a size category without much effort, it is time to take advantage of the many field marks that shape and posture offer.

Posture
While there is nothing noteworthy about the normal posture of most birds, there are two postures—horizontal and vertical—that are distinctive, and single out a few groups or species. Some birds, generally those with short legs and streamlined bodies, have a horizontal posture. Birds that show this posture, like the nightjars and nighthawks, the terns, and the Black Skimmer, usually obtain most of their food while flying; their short legs and streamlined bodies reflect their aerial feeding habits, for these birds are built to move swiftly through the air, and make little use of their feet. Another equally unusual posture is vertical. Some birds, such as the raptors and owls, perch in an upright position for long periods as they scan their surroundings for prey. A vertical posture is also seen in woodpeckers, birds that cling to the trunks of trees, and in some diving birds whose feet are so far back on the body that they must hold themselves upright to maintain their balance.

Shape
When looking at body shapes of birds, it is easy to single out those that are stocky and slender, since most species have neither of these shapes, but are somewhere in between. Stocky species have plump bodies, and are found in every size group; they look round, whether they are well fed or not. Slender birds have thin bodies and often have long tails. Their thin appearance has nothing to do with how much they have eaten; their slenderness serves as a reliable field mark at all times.

Birds also vary in the shape of the head and neck, the bill, the wings, the tail, and the legs. These different shapes all represent adaptations to conditions in each species' habitats. Of all the parts of a bird, the bill is the most variable. This is because every species uses its bill to carry out its own specialized feeding methods; a bill functions as part of a bird's mouth, but also has to perform many of the activities that in humans would be carried out by the hands. This becomes particularly obvious when you compare birds that feed on or near water with those that search for food in meadows, shrubs, or trees.

Horizontal   Vertical   Stocky   Slender

As you make this comparison, you will begin to understand the distinction birders make between water birds and land birds. Spear-shaped bills, for example, are a hallmark of water birds that catch fish by diving or spearing them. Long and strongly downcurved bills are found in water birds that probe in sand, mud, or shallow water. It is easy to recognize a duck by its distinctive bill, and ducks are obviously water birds. Among land birds, the birds of prey have hooked bills for tearing apart their victims; seed-eating birds have bills that are conical, useful for crushing the tough coats of seeds; and birds that forage for insects in bushes and trees have slender bills, all they need to "handle" this softer food.

For birds in flight, it is the shape of the wings and of the tail that will help you identify a species or group. Broad, rounded wings are typical of many birds that soar over open country or mountains—like many hawks and eagles. Wings that are long, narrow, and pointed belong to birds like shearwaters, gannets, and boobies that soar or glide over the sea, because these birds must be agile enough to change course the moment the wind bounces suddenly off a wave. Tails are equally prominent field marks for birds in flight or on the ground. Most terns can be distinguished from gulls by their forked tails. Among swallows, a bird with a forked tail must be a Barn Swallow, and one with a square-tipped tail must be a Cliff Swallow.

Arrangement of the Illustrations

In the following pages, I have organized the text and illustrations into several categories: one dealing with overall shape and posture, and then special sections on the parts of a bird—the head and neck, the bill, the wings, and the tail. By concentrating in turn on heads, necks, bills, wings, and tails, we are in a sense taking a bird apart and examining the variation in each part.

Using Shape and Posture in the Field

In each of the preceding chapters I have urged you to take careful notes in the field. By keeping notes on what you have seen, you have gradually built up a picture of the bird and have collected detailed information on habitat, size, and behavior. At each stage you have eliminated many birds and have narrowed your choices to ever fewer groups and species. You should now add a new heading, "Shape and Posture," to your notes and carefully describe each part of a bird you see. This will not only help to fill out your impression of the bird, but should also complete the process of eliminating groups. If you haven't been able to pinpoint a group before now, the odds are that seeing a conical bill, or a forked tail, or short, broad, rounded wings will enable you to identify the group to which a bird belongs, and in many cases, to name the species.

Shape and Posture

Birds come in a variety of shapes, but those at the extremes—slender or stocky—are noticeable enough to be helpful field marks.

A feature closely related to the overall shape of a bird is the posture it habitually assumes.

Slender

A truly slender bird is one with a slender body, like that of a heron or a cuckoo, but the impression of slenderness can also be created by a long tail or a long neck. Townsend's Solitaire has a body that is nearly the same as other thrushes, but its long tail makes it appear slender.

Tricolored Heron

Stocky

Some birds are plump. Others are not as heavy-bodied, but appear stocky because they have a large head, short neck, or short tail.

Ruffed Grouse

Horizontal

Of the birds that have a strikingly horizontal posture, the terns, the Black Skimmer, and the nightjars are slender, with very short legs and small feet. Shrikes have a more normal build, but they have a habit of holding the tail straight out behind them, so that, when perching, shrikes also have a horizontal appearance.

Common Nighthawk

Vertical

Among the birds with an upright posture, some are divers whose legs are placed far back on the body to give them forward thrust underwater. Others spend their time perched in trees watching for moving prey. Woodpeckers and the Brown Creeper cling to the trunks of trees in an upright position, propped up by their tails.

Short-eared Owl

While there is nothing that is diagnostic about the posture of most birds, some species or groups characteristically assume a distinctive horizontal or vertical posture that is a valuable clue in identification.

Greater Roadrunner

Yellow-billed Cuckoo

Cactus Wren

Yellow-crowned Night-Heron

Black Skimmer

Caspian Tern

Downy Woodpecker

Rough-legged Hawk

Head and Neck

The size of the head, its posture, and a variety of features, such as crests, ear tufts, combs, and frontal shields, all form part of what we see when we look at an unfamiliar bird, and sometimes can be the deciding factor in determining the species.

Head

Small

Combs or Crest

Combs

Shield or Plumes

Shield

Neck

Short

The neck, too, can be unusually long or short, bear plumes, or be held in any of a number of distinctive postures.

Large

Naked

Bushy Crest

Pointed Crest

Ear Tufts or Facial Plumes

Head or Neck Plumes

Long

Arched

Bills
Water Birds

Water birds show great variation in bill shape. Species with long, slim bills use them for probing mud or sand in search of food, or for snatching food from the surface like a pair of tweezers. Pointed or spear-shaped bills are used for seizing fish or actually spearing them. In general, the stouter the bill, the larger the prey the bird can capture.

Long and Stout

Very Large

Long and Slim

Typical

Slim

Pointed

Spear-shaped

Stout

Unusual bill shapes reflect specialized methods of feeding. The largest bills of any birds are those of the pelicans, used for scooping up fish.

The American Avocet sweeps its upcurved bill from side to side, straining small animals out of the water. The Black Skimmer trails its long lower mandible in the water as it flies close to the surface; the instant the mandible touches a fish, the bill snaps shut like a trap, capturing the prey.

Slightly Downcurved

Strongly Downcurved

Upcurved

Downcurved

Needlelike

Slightly Downcurved

Slender

Long Lower Mandible

Bills
Water Birds

Water birds with shorter bills range from the familiar ducks and geese to the puffins, which use their odd, parrotlike bills for seizing fish during underwater dives. The bill of the aptly named Roseate Spoonbill is used to strain food from the water.

Short and Stout

Parrotlike

Flat

Knobbed

Ducklike

Short

Gooselike

Short

Gulls have a distinctive, all-purpose bill that is useful for scavenging, probing, tearing at captured prey, or stealing food from other birds.

Gull-like

Chickenlike

Shovel-shaped

Spoon-shaped

Typical

Long

Long

Knobbed

Bills
Land Birds

Land birds do not show as much variation in bill shape as water birds, and the bills of land birds are often less conspicuous. Nevertheless, they are very helpful as field marks, especially for identifying the group to which a species belongs. The variation also reveals something about the feeding habits of the birds.

Short

Very Small

Long

Very Long

Hooked

Typical

Stout

Typical

The shortest bills are those of nighthawks, nightjars, swifts, and swallows, which capture insects in flight.

Slender bills usually indicate that a bird feeds on insects; birds with short or stout conical bills use them to crack open seeds.

The very long bill of a hummingbird is used to probe flowers. A typical hooked bill is a hallmark of raptors; chickenlike bills are useful for picking seeds or insects from the ground.

Conical

Slender

Long and Pointed

Strongly Downcurved

Parrotlike

Crossed

Conical

Chickenlike

Wings and Tails in Flight

If you see an unfamiliar bird in flight, note the shape of the wings and tail to help you identify the species. In most birds, the shape of the wings and tail reflect the style of flight. The wings drive the birds forward, while the tail serves as rudder for steering as the bird moves through the air.

Long and Narrow Wings

Rounded

Broad and Rounded Wings

Short

Short Tail

Fan-shaped

Long Tail

Square-tipped

Rounded wings and fan-shaped tails are usually found in birds that soar, and pointed wings are often associated with speed or with long and sustained wingbeats.

Long or forked tails, as well as tails with long central feathers, give birds great maneuverability, and are found in birds that are very agile on the wing, such as terns and frigatebirds.

Pointed

Pointed and Kinked

Long

Long and Kinked

Shallowly Forked

Pointed

Forked

Long Central Feather

Tails

In perching birds or those that are standing on the ground or resting on the water, the length of the tail contributes to the overall impression the bird makes. A short tail can make a bird look stocky, and a long tail is associated with a slender overall shape.

Short Tail

Not Visible

Long Tail

Long

Tail Shapes

Fan-shaped

Tail Tip

Notched

The shape of the tail is often useful in distinguishing species, and may be fan-shaped, pointed, or keeled. In a few species, the central tail feathers are much longer than the rest.

The tip of the tail may be notched, square-tipped, or rounded. The rounded tail tips of some birds are often visible when the bird is perched, but the tail feathers are sometimes more obvious when the bird is in flight.

Very Short

Short

Very Long

Long Central Feather

Pointed

Keeled

Square-tipped

Rounded

How to Use the Shape and Posture Charts

The nine shape and posture charts are divided into two categories, just like the behavior charts, depending on whether the bird was seen on or near water, or on or over land. Additionally, on land, birds in the three smallest categories are grouped according to whether they forage in bushes and in trees, or on the ground.

Shape and posture are the two field marks that often relate a bird most closely to its habitat. Different shapes and postures are adaptations that enable birds to feed and nest successfully in their preferred environment. In this chapter, I have focused on those groups and species where the relationship of shape and posture to habitat is most noticeable.

By now you have already observed some birds and taken notes about shape and posture. I suggest you compare your notes with the charts I have prepared and narrow your choices to a group or a species. As a further aid in mastering this field mark category, you should study the preceding pages and become familiar with the different overall postures and shapes of birds, and with differences in wings, tails, heads, and bills. When you are confident that you can spot a bird's salient features, you are ready to go into the field.

Tips for the Field

For more practice working with shape and posture, I suggest that you follow the same procedure as you did when you were learning about bird behavior. Pick a habitat, and after you have noted size and behavior, study each part of each bird you see. First, look at its overall shape, and decide whether the bird strikes you as slender, stocky, or neither. Then note the bird's posture. Is it vertical, horizontal, or neither? Next you should carefully check each part—the head, neck, bill, wings, tail, and legs. Refer to the names of the parts illustrated in the photograph section and listed on the charts. Try to form mental images of the parts—a conical bill, rounded wings, a forked tail. If the bird flies while you are watching it, note the shape of the wings and tail, because distinctive wing and tail shapes can usually be seen most clearly when a bird is in flight. Birders sometimes deliberately flush a bird in order to get a good look at these features. As you learn the different shapes of birds, remember that shape and behavior go hand in hand. Long, slender bills are usually found in birds that probe, whether they are shorebirds feeding on a beach or mudflat, or hummingbirds seeking nectar at flowers. Long legs are characteristic of birds that wade. By making correlations between shape and behavior, you will find it easier to remember both of these kinds of field marks.

Arrangement of Shape and Posture Field Marks

Because of the close relationship between shape and posture, and behavior, I have organized the nine shape and posture charts according to the same overall plan used in the behavior charts, by water and land. The different shapes and postures are listed in the column on the left, and the bird groups and species are arranged across the top. On each chart, birds with similar shapes and postures are placed next to each other. You should remember that not all birds found in water habitats are water birds and vice versa. Yet, the terms land birds and water birds are used frequently by birders, and you should get to know which birds belong to these broad groups. For this reason, on the water charts I have indicated land birds with a black dot, and in the charts dealing with land habitats, the black dot indicates water birds.

On or near Water
Sparrow-sized and Robin-sized
Pigeon-sized and Crow-sized,
two charts
Goose-sized and Very Large

On or over Land
Very Small and Sparrow-sized
Sparrow-sized and Robin-sized
Robin-sized
Pigeon-sized
Crow-sized and Goose-sized, and
Very Large

Using the Charts

As you look at the charts, the first shapes and postures listed are
those that apply to the whole bird, and then the variations in each
part of the birds are listed in order from head to tail. You will also
note that long legs and short wings are listed on the charts, although
they were not illustrated in the photo section, because these are
obvious features. Wedge-shaped tail has also been added to help you
distinguish ravens, which have a wedge- or V-shaped tail, from crows,
which have a rounded tail.

When you have seen a bird, and know its habitat type and size, select
the appropriate chart and then look for a distinctive shape—head, bill,
wings, or tail. Don't forget the specific habitat and size categories for
each group or species are listed at the bottom of the chart. The red
square in the chart indicates "giveaway" features, like the strongly
downcurved bills of the Long-billed Curlew and Whimbrel on the third
chart.

Verifying Your Identifications with the Field Guide

Now that you have collected notes on a bird's habitat, size, behavior,
and shape and posture, you are ready to use your field guide to
confirm an identification or to distinguish between very similar
species. The shape and posture charts will tell you that the crow-sized
shorebird you saw at the beach, probing in the sand with a strongly
downcurved bill, was either a Long-billed Curlew or a Whimbrel. If
you now refer to your companion field guide, where these two similar
species are pictured on facing pages, you will learn at once that the
bird had to be a Long-billed Curlew, because its very long bill
matches that of the bird you saw at the beach.

If you visit brushy open country or a thicket, and see a robin-sized
gray bird with a crest and a conical bill foraging in bushes, the shape
and posture charts will tell you that the bird could only have been one
of two species, a Northern Cardinal or a Pyrrhuloxia. When you
consult your field guide, you need only look up these two birds. You
already know that the male Northern Cardinal is bright red, and will
discover in the field guide that the Pyrrhuloxia is mainly gray, with
red in the wings and tail; the bird you saw was gray, and so you know
it must have been a Pyrrhuloxia. In this case, color was the deciding
factor. It is time to turn to the field-mark category of color and
pattern, the final step in breaking groups down into species, and
identifying birds in the field.

Shape and Posture 1
On or near Water

Sparrow- and Robin-sized
- ■ *Giveaway Field Mark*
- ● *Land Birds*

1. Least Grebe
2. Auks
3. Little Gull
4. Black Tern
5. Least Tern
- ● 6. Swallows
- ● 7. Barn Swallow
- ● 8. Cliff Swallow

	1	2	3	4	5	6	7	8
Slender				■	■			
Stocky		■						
Horizontal				■	■			
Head: Small								
Large head		■						
Bushy crest								
Shield								
Short neck								
Bill: Long, slim								
Pointed, slim				■	■			
Needlelike								
Slightly downcurved, slim								
Spear-shaped, stout								
Spear-shaped, slender	■							
Gull-like			■					
Chickenlike								
Very small						■	■	■
Slender, short								
Wings: Long, narrow, pointed			■	■	■	■	■	■
Short, broad, rounded								
Short		■						
Tail in Flight: Fan-shaped, short			■					
Shallowly forked, short				■	■			
Forked, long							■	
Tail: Not visible	■							
Very short								
Tail Tip: Notched						■		
Square								■
Legs: Long								
Saltwater Habitats		■	■	■	■	■	■	
Freshwater Habitats	■		■	■	■	■	■	■
Sparrow-sized						■	■	■
Robin-sized	■	■	■	■	■			

9. Plovers
10. Red-necked and
    Wilson's phalaropes
11. Dowitchers and Stilt Sandpiper
12. Lesser Yellowlegs
13. Northern Jacana
14. Dunlin
15. Common Snipe
16. American Woodcock
17. Black and Yellow rails

18. Virginia Rail
19. Sora
● 20. American Dipper
● 21. Waterthrushes
22. Green Kingfisher

Shape and Posture 2
On or near Water

Pigeon- and Crow-sized
Grebes, Auks, Ducks, and Geese
■ *Giveaway Field Mark*

1. Pied-billed Grebe
2. Red-necked Grebe
3. Western Grebe
4. Eared Grebe
5. Horned Grebe
6. Rhinoceros Auklet
7. Tufted Puffin
8. Puffins

	1	2	3	4	5	6	7	8
Slender			■					
Stocky	■					■	■	■
Vertical								
Head: Small			■	■	■			
Large head								
Bushy crest								
Pointed crest								
Shield								
Ear tufts/Facial plumes				■	■	■	■	
Short neck								
Long neck			■					
Bill: Spear-shaped, slender		■	■	■	■			
Parrotlike							■	■
Chickenlike	■							
Knobbed, flat								
Shovel-shaped								
Ducklike, short								
Ducklike, typical								
Ducklike, long								
Gooselike, short								
Wings: Short						■	■	■
Tail in Flight: Long central feathers								
Tail: Not visible	■	■	■	■	■			
Legs: Long								
Saltwater Habitats	■	■	■	■	■	■	■	■
Freshwater Habitats	■	■	■	■	■			
Forest Habitats								
Open Habitats								
Pigeon-sized	■			■	■	■	■	■
Crow-sized		■	■					

9. Guillemots
10. Razorbill
11. Murres
12. Ducks
13. Whistling-ducks
14. Wood Duck
15. Northern Shoveler
16. Northern Pintail
17. Mergansers
18. Ruddy and Masked ducks
19. Bay and Sea Ducks
20. Oldsquaw
21. Canvasback
22. Bufflehead and Goldeneyes
23. Tufted Duck
24. King Eider
25. Scoters
26. Ross' Goose

9	10	11	12	13	14	15	16	17	18	19	20	21	22	23	24	25	26

Shape and Posture 3
On or near Water

*Pigeon- and Crow-sized
Gulls, Terns, Shorebirds, and
Others*
■ *Giveaway Field Mark*
● *Land Birds*

1. Tropicbirds
2. Jaegers
3. Shearwaters
4. Gulls and Skuas
5. Sabine's Gull
6. Terns
7. Gull-billed Tern
8. Sandwich Tern

	1	2	3	4	5	6	7	8
Slender						■	■	■
Horizontal						■	■	■
Head: Small								
Large head								
Bushy crest								■
Shield								
Long neck								
Bill: Long, slim								
Upcurved, long, slim								
Downcurved, long, slim								
Pointed, slim						■		■
Spear-shaped, stout	■						■	
Long lower mandible								
Gull-like		■		■	■			
Chickenlike								
Hooked								
Wings: Long, narrow, rounded								
Long, narrow, pointed	■	■	■	■	■	■	■	■
Long, broad, rounded								
Long, rounded, kinked								
Tail in Flight: Fan-shaped, short				■	■			
Shallowly forked, short					■		■	
Pointed, short								
Forked, long						■		■
Long central feathers	■	■						
Tail: Very short								
Long								
Legs: Long								
Saltwater Habitats	■	■	■	■	■	■	■	■
Freshwater Habitats		■		■		■		
Open Habitats				■			■	
Pigeon-sized		■	■	■	■	■	■	■
Crow-sized	■	■	■	■				

9. Brown Noddy
10. Royal, Caspian, and Elegant terns
11. Black Skimmer
12. Black-necked Stilt and Willet
13. Long-billed Curlew and Whimbrel
14. American Avocet and Godwits
15. Greater Yellowlegs
16. Oystercatchers
17. Rails
18. Purple Gallinule
19. Common Moorhen and American Coot
20. Least Bittern
21. Herons and Egrets
● 22. Osprey
● 23. Snail Kite
● 24. American Swallow-tailed Kite
● 25. Northern Harrier
26. Belted Kingfisher

| 9 | 10 | 11 | 12 | 13 | 14 | 15 | 16 | 17 | 18 | 19 | 20 | 21 | 22 | 23 | 24 | 25 | 26 |

Shape and Posture 4
On or near Water

Goose-sized and Very Large
■ *Giveaway Field Mark*
● *Land Birds*

1. Magnificent Frigatebird
2. Albatrosses
3. Boobies and Northern Gannet
4. Gulls
5. Pelicans
6. Loons
7. Red-throated Loon
8. Cormorants

	1	2	3	4	5	6	7	8
Slender	■							
Stocky								
Vertical								■
Head: Bushy crest								
Long neck								■
Arched neck								
Bill: Very large, long, stout					■			
Slightly downcurved, long, stout								
Strongly downcurved, long, stout								
Spear-shaped, stout			■			■		
Spear-shaped, slender							■	
Gull-like				■				
Spoon-shaped								
Ducklike, long								
Gooselike, short								
Gooselike, long								
Gooselike, knobbed								
Hooked								
Wings: Long, narrow, rounded								
Long, narrow, pointed		■	■	■				
Long, narrow, pointed, kinked	■							
Short, broad, rounded								
Long, broad, rounded								
Tail in Flight: Fan-shaped, short				■				
Pointed, short			■					
Forked, long	■							
Tail: Very short						■	■	
Long and fan-shaped								
Legs: Long								
Saltwater Habitats	■	■	■	■	■	■	■	■
Freshwater Habitats			■	■	■	■	■	■
Goose-sized	■		■	■		■	■	■
Very Large		■	■		■	■		■

205

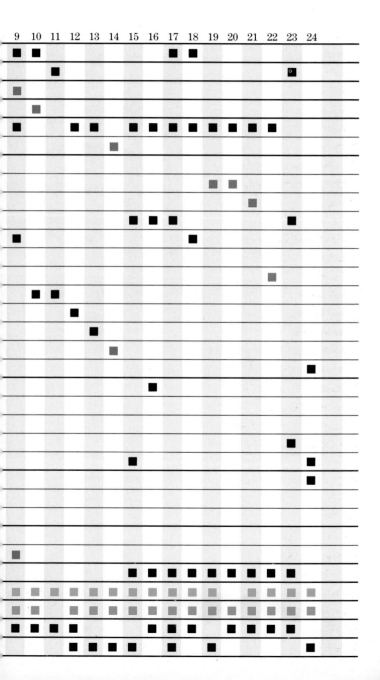

9. Anhinga
10. Common Merganser
11. Common Eider
12. Geese
13. Swans
14. Mute Swan
15. Cranes
16. American Bittern
17. Herons and Egrets
18. Snowy Egret
19. Wood Stork
20. Limpkin
21. Ibises
22. Roseate Spoonbill
23. Night-herons
● 24. Bald Eagle

Shape and Posture 5
On or over Land

Very Small and Sparrow-sized
Foragers in Bushes and in Trees
■ *Giveaway Field Mark*

1. Woodpeckers
2. Owls
3. Flycatchers
4. Wrens
5. Nuthatches
6. Titmice
7. Cedar Waxwing
8. Red-whiskered Bulbul

	1	2	3	4	5	6	7	8
Slender								
Stocky		■		■	■			
Vertical	■	■	■					
Head: Large		■	■	■	■	■		
Pointed crest						■	■	■
Short neck				■	■	■		
Bill: Very small						■		
Conical, short								
Slender, short			■					
Very long								
Long, pointed					■			
Slightly downcurved				■				
Parrotlike								
Crossed								
Stout								
Conical, stout								
Wings: Long, narrow, pointed								
Tail in Flight: Forked, long								
Tail: Short				■	■		■	
Long								
Pointed								
Tail Tip: Notched								
Square								
Saltwater Habitats				■				
Freshwater Habitats			■	■				
Forest Habitats	■	■	■	■	■	■	■	
Open Habitats	■	■	■	■	■		■	
Urban and Residential Habitats	■		■	■		■	■	■
Very Small				■	■			
Sparrow-sized	■	■	■	■	■	■	■	■

9. Swifts	18. Warblers
10. Swallows	19. Yellow-breasted Chat
11. Cliff Swallow	20. Western Tanager
12. Barn Swallow	21. Blue Grosbeak
13. Kinglets	22. Bobolink
14. Bushtit	23. Finches
15. Wrentit	24. Crossbills
16. Gnatcatchers	25. Budgerigar
17. Vireos	26. Hummingbirds

1. Horned Lark
2. Eurasian Skylark
3. Plovers
● 4. American Woodcock
5. Quails
6. Scaled Quail and Northern Bobwhite
7. Montezuma Quail
8. Common Ground-Dove

Sparrow- and Robin-sized
Ground Foragers
■ *Giveaway Field Mark*
● *Water Birds*

	1	2	3	4	5	6	7	8
Slender								
Stocky				■	■	■	■	
Head: Small					■	■	■	■
Large head			■					
Bushy crest							■	
Pointed crest		■			■			
Ear tufts	■							
Head plumes					■			
Short neck			■	■				
Bill: Pointed, slim				■				
Conical, short								
Slender, short								
Long, pointed								
Slightly downcurved								
Strongly downcurved								
Stout								
Chickenlike					■	■	■	
Wings: Long, narrow, pointed			■					
Short, broad, rounded				■	■	■	■	
Tail: Short								
Long								
Fan-shaped								■
Keeled								
Tail Tip: Square								
Rounded								
Saltwater Habitats	■		■					
Freshwater Habitats			■	■				
Forest Habitats				■	■	■	■	■
Open Habitats	■	■	■		■	■	■	■
Urban and Residential Habitats					■			
Sparrow-sized	■	■						■
Robin-sized				■	■	■	■	■

9. Inca Dove
10. European Starling
11. Meadowlarks
12. Common Grackle
13. Cowbirds
14. Buntings and Dickcissel
15. Sparrows
16. Juncos
17. Longspurs
18. Towhees
19. Green-tailed Towhee
20. Northern Mockingbird
21. American Robin
22. Pipits
23. Thrashers
24. Bendire's and Brown thrashers
25. Sage Thrasher

Shape and Posture 7
On or over Land

Robin-sized
Foragers in Bushes and in Trees
■ *Giveaway Field Mark*

1. Doves
2. Mourning Dove
3. Ringed Turtle-Dove and Spotted Dove
4. American Kestrel
5. Shrikes
6. Nightjars
7. Nighthawks
8. Common Poorwill

	1	2	3	4	5	6	7	8
Slender						■	■	
Stocky								
Horizontal					■	■	■	■
Vertical								
Head: Small	■	■	■					
Large head					■	■	■	■
Bushy crest								
Pointed crest								
Ear tufts								
Short neck					■			
Bill: Very small						■	■	■
Slender, short								
Long, pointed								
Slightly downcurved								
Hooked				■	■			
Stout								
Conical, stout								
Wings: Long, narrow, pointed							■	
Tail in Flight: Forked, long								
Tail: Short								■
Long				■	■	■	■	
Fan-shaped	■							
Pointed		■						
Tail Tip: Notched							■	
Square								
Rounded				■	■	■		■
Forest Habitats	■	■				■	■	■
Open Habitats								
Urban and Residential Habitats	■	■	■	■			■	
Robin-sized	■	■	■	■	■	■	■	■

9. Cuckoos
10. Townsend's Solitaire
11. Purple Martin
12. Cactus Wren
13. Owls
14. Flammulated Owl and
    Screech-owls
15. Flycatchers
16. Scissor-tailed Flycatcher
17. Woodpeckers

18. Crested Myna
19. Northern Cardinal and
    Pyrrhuloxia
20. Bohemian Waxwing and
    Phainopepla
21. Jays
22. Blue and Steller's jays
23. Pinyon Jay
24. Tanagers
25. Grosbeaks
26. Gray Catbird

Pigeon-sized
- Giveaway Field Mark
- Water Birds

● 1. Upland Sandpiper
2. Pigeons and Doves
3. Gray Partridge
4. White-tailed Ptarmigan
5. Chukar
6. Owls
7. Northern Flicker
8. Elegant Trogon

	1	2	3	4	5	6	7	8
Slender								
Stocky			■	■	■	■		
Horizontal								
Vertical						■	■	■
Head: Small	■	■	■	■				
Large head					■	■		■
Combs					■			
Short neck								■
Long neck	■							
Bill: Very small								
Long, pointed								
Hooked								
Parrotlike								
Stout							■	
Chickenlike			■	■	■			
Wings: Long, narrow, pointed								
Short, broad, rounded			■	■	■			
Tail in Flight: Square-tipped, long								
Tail: Long								■
Fan-shaped	■	■	■	■				
Pointed								
Keeled								
Tail Tip: Square								■
Rounded								
Saltwater Habitats		■						
Freshwater Habitats		■						
Forest Habitats		■		■		■	■	■
Open Habitats	■	■	■	■	■	■	■	
Urban and Residential Habitats		■					■	
Pigeon-sized	■	■	■	■	■	■	■	■

9. Clark's Nutcracker
10. Chuck-will's-widow
11. Grackles
12. Magpies
13. Brown Jay
14. Anis
15. Merlin
16. Sharp-shinned Hawk

Shape and Posture 9
On or over Land

*Crow- and Goose-sized, and
Very Large*
■ Giveaway Field Mark

1. Pileated Woodpecker
2. Raptors
3. Golden Eagle
4. Black Vulture and California Condor
5. Turkey Vulture
6. Northern Harrier
7. Black-shouldered and Mississippi kites
8. Falcons

	1	2	3	4	5	6	7	8
Slender						■		
Stocky								
Vertical	■	■	■				■	■
Head: Small					■	■		
Large head								
Naked head				■	■			
Combs								
Bushy crest								
Pointed crest	■							
Ear tufts								
Head or neck plumes								
Long neck	■							
Bill: Hooked		■	■	■	■	■	■	■
Stout								
Chickenlike								
Wings: Long, narrow, rounded					■	■		
Long, narrow, pointed							■	■
Short, broad, rounded								
Long, broad, rounded		■	■	■				
Tail in Flight: Fan-shaped, short		■	■	■				
Square-tipped, long							■	
Wedge-shaped								
Tail: Long					■	■		■
Fan-shaped								
Pointed								
Saltwater Habitats		■				■	■	■
Freshwater Habitats		■				■	■	■
Forest Habitats	■	■	■	■	■	■		
Open Habitats	■	■	■	■	■	■	■	■
Urban and Residential Habitats								■
Crow-sized	■	■					■	■
Goose-sized				■	■			
Very Large				■	■			

9. Crested Caracara
10. Owls
11. Long-eared and
    Great Horned owls
12. Northern Goshawk and
    Cooper's Hawk
13. Black Francolin
14. Ptarmigans
15. Blue and Spruce grouse
16. Sharp-tailed Grouse
17. Sage Grouse

18. Ruffed Grouse
19. Prairie-chickens
20. Wild Turkey
21. Ring-necked Pheasant
22. Plain Chachalaca
23. Greater Roadrunner
24. Ravens
25. Crows

Color and Pattern

Color and Pattern

Common bird species are covered in nine color and pattern charts at the end of this chapter.

Birds have a keen sense of sight and, except for nocturnal species, they can distinguish colors. This is why we find such a wide variety of colors and patterns in birds. Colors and patterns, like behaviors and different shapes and postures, are adaptations that serve useful functions in the life of any bird. Bright or conspicuous colors enable members of a particular species to recognize one another. For breeding males, the colors serve to warn other males of the same species to stay away. These same colors and patterns attract females of the species, and the two birds soon form a pair. Conspicuous colors are also valuable to gregarious birds, such as crows or egrets, which must be highly visible to one another as they gather at roosts or assemble at rich feeding areas. On the other hand, dull colors like brown, gray, or olive often serve as camouflage, enabling the birds that wear them to avoid predators. Females are also often clad in concealing colors, so that they do not attract attention to nests or young.

We respond as quickly to a bird's color and pattern as they do. But the colors and patterns of North American birds are so varied that even though they may be very obvious to us, sorting out all of our birds on the basis of color alone would be an almost impossible task. Instead, I have stressed that you should consider habitat, size, behavior, and shape and posture first before turning to color and pattern. Expert birders take note of these features automatically, and when they use color and pattern, only a few choices remain.

Color

The colors of birds are not as simple as they might at first appear. The major colors found in our birds are blue, black, white, orange, rufous, yellow, olive, gray, brown, and red. Most of these colors come in many different shades, and in some birds, they are glossy or iridescent. How you see a color often depends on lighting conditions. Gray birds may look olive in light reflected from nearby green leaves, while olive birds may appear gray on dull days or at times of the year when their plumage is worn and faded. On cloudy days, the blue of an Indigo Bunting may appear black. Similarly, the angle of light may make the iridescent throat of a hummingbird appear black. To add to the confusion, the names of birds often describe colors very loosely and even inaccurately: For example, the Red-tailed Hawk has a rufous tail, and the American Black Duck is actually brown.

Pattern

While some birds are entirely of one color, most show combinations of colors that form patterns. These patterns occur in great variety, but certain parts of a bird's body tend to contribute to its overall pattern; among these are the crown, eyebrows, throat, breast, wing bars, rump, and outer tail feathers. Additionally, the upperparts or underparts may be streaked, spotted, or barred.

Except in species that are extremely similar, the features that are the most conspicuous provide the most useful field marks. These are usually the more brightly colored patterns that contrast boldly with other color markings. Thus, a bird may have a crown of one color and eyebrows of another; if the color of the eyebrow is brighter than the

The first chart is devoted to ducks, one of the most familiar groups of birds. Woodpeckers, another distinctive and easily recognizable group, are featured in the last chart.

The large group of common songbirds is covered in seven charts, according to basic colors: blue, black, orange and red, yellow, olive, gray, and brown.

crown and contrasts with the rest of the bird, the eyebrow is a better field mark than the crown.

Arrangement of the Illustrations

When I talk about a bird's color and pattern as a field mark, I am referring to parts of a bird's body that sport a brighter or more obvious color than the rest of it. The purpose of the color section in this chapter and the accompanying text is to familiarize you with the terms used to describe a particular part of the body and to teach you how to see color and pattern field marks on birds. To simplify this task, I have included only common species of three groups—the songbirds, which are generally small land birds, the woodpeckers, and the ducks. When you read descriptions in your field guide, you will find these basic definitions useful.

As a group, songbirds show the range of colors birds come in, from bright blue and red to dull olive and gray. I have assigned birds to a color category based on the color that is most striking. As a result, you will find a bird that is mostly blue with the blue group. For birds that have more than one dominant color, I have chosen the most conspicuous. The Yellow-headed Blackbird, for instance, is largely black, but has a bright yellow head and upper breast, its most noticeable feature, so I have included it among the birds whose most striking color is yellow. The American Robin has a dark gray back, but the most distinctive color of this species is its rufous breast; so this familiar species is placed among the orange and rufous birds.

In contrast to the multicolored songbirds, woodpeckers are mostly black-and-white. When studying this group, you will appreciate the importance of their red or yellow field marks in distinguishing the species from each other.

The third group featured is ducks, which, like the songbirds, display a range of colors. Because ducks are larger birds and can be more readily observed in the field, I suggest you watch them and note where color occurs on their bodies. Become acquainted with such terms as ear patch, cheek, and shield. You will be able to apply this knowledge to other birds. Look over these pages carefully until you are familiar with the names of the various parts of a bird and can add them to your vocabulary. To assist you in learning these terms, we have included a drawing with each set of photographs, showing some of the most important pattern elements for each group of species.

Using Color and Pattern in the Field

After studying the illustrations and text, go into the field, find a bird, and try to assign it to a major color. Then jot down other colors and patterns on the bird. Don't worry about whether each feature you see is a field mark, but try to become accustomed to seeing all of the parts of a bird, and use this opportunity to test your recollection of the terms described and pictured here. After trying this a few times, you will find that you can take one quick glance at a bird and remember nearly every detail. You will be able to bring home a complete and accurate mental image of the bird, whether it is one you already know, or a species that is new to you.

Ducks

Using color and pattern, you can identify most ducks to species. The first group shown here includes the most brightly colored ducks. Important field marks to note are the color of the head, forehead, ear patch, face patch, cheeks, breast, flanks, and rear end.

Green Head

These three ducks have a green head. The Mallard and Northern Shoveler are very similar, but the Mallard has a chestnut breast and gray flanks, while the Northern Shoveler has a white breast and chestnut flanks. The Northern Shoveler also has a large black bill. With its pointed crest, bold head pattern, buff flanks, and long tail, the Wood Duck is unmistakable.

Northern Shoveler

Green Ear Patch, Pale Forehead

Both the American Wigeon and Green-winged Teal have a glossy green ear patch, but the rest of the head in the Green-winged Teal is mainly rufous, like the head of the Eurasian Wigeon. The wigeons have pale foreheads—white in the American and buff in the Eurasian. The smaller Green-winged Teal lacks this pale forehead.

American Wigeon

Brown or Rufous

The Cinnamon Teal is wholly chestnut or rufous, except for its two blue wing patches, most easily seen in flight. Although the Gadwall is mainly gray, it has a sandy brown head. The slender Northern Pintail is also mainly gray, but has a rich brown head and a distinctive white neck stripe and breast.

Cinnamon Teal

White Face Patch or Cheek

The Blue-winged Teal, Ruddy Duck, and Harlequin Duck all have conspicuous white face patches. In the Blue-winged Teal, the patch forms a white crescent at the base of the bill. The Ruddy has bold white cheeks, while the seagoing Harlequin Duck has a white triangle in front of the eye, as well as smaller white spots on the side of the head.

Blue-winged Teal

Forehead | Ear patch
Breast | Rear end
Flank

Mallard

Wood Duck

Green-winged Teal

Eurasian Wigeon

Gadwall

Northern Pintail

Ruddy Duck

Harlequin Duck

Ducks

Most of these ducks are boldly patterned in black and white. Key field marks are the color of the head or face patch, the flanks, and the bill or shield.

White Head Patch or Face Patch
Both the Hooded Merganser and Bufflehead have a large white patch on the side of the head. In the Hooded Merganser, this patch is bordered with black, and can be erected like a fan. The patch of the Bufflehead extends around the back of the head, and lacks the merganser's black border. Barrow's Goldeneye has a smaller white patch in front of the eye.

Hooded Merganser

Black-and-White Pattern
Although the heads of the Red-breasted and Common mergansers are glossy green, they usually look black at a distance. As a result, these slender ducks appear black-and-white in the field. The Oldsquaw also looks black-and-white, but unlike the other two, its head is largely white.

Red-breasted Merganser

White Sides or Body
The Ring-necked Duck, Greater Scaup, and Canvasback are all bay ducks with black breasts, black rear ends, and white or pale gray flanks. The Ring-necked Duck is the only one with a black back and a boldly patterned bill. The Canvasback has a rufous head and a long black bill, in contrast to the glossy greenish-black head and bluish bill of the Greater Scaup.

Ring-necked Duck

Shield or Bill
Both the Common and King eiders have a frontal shield. The bright orange shield of the King is a good field mark, but that of the Common is small, and often not evident. These two eiders are better separated by the color of the back: black in the King and white in the Common. The Surf Scoter has a multicolored, swollen bill with a small knob.

Common Eider

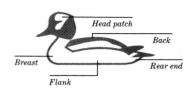

Head patch

Back

Breast

Rear end

Flank

Bufflehead

Barrow's Goldeneye

Common Merganser

Oldsquaw (winter)

Greater Scaup

Canvasback

King Eider

Surf Scoter

Blue

The overall color, plus that of the underparts and back, are important field marks for this group. Also note the color and pattern on the face or throat.

Deep Blue
Each of these sparrow-sized birds has deep blue somewhere in its plumage. Unmistakable, the Painted Bunting has a blue head, with green upperparts and red underparts. The Blue Grosbeak is largely deep blue, but has two rufous wing bars. Although wholly deep blue, the Indigo Bunting often appears black in poor light.

Painted Bunting

Blue with a Rufous Breast
Three birds are largely blue with a rufous breast. The Lazuli Bunting is the only one of these that has two white bars and a conical bill. The two bluebirds have ranges that are mainly separate, but where they occur together they can be distinguished by the rufous back and blue throat of the Western; the Eastern Bluebird has an all-blue back and rufous throat.

Lazuli Bunting

Solid Blue Overall
The Mountain Bluebird and the Pinyon Jay are both solid blue, but differ in size and shape. The Mountain Bluebird is smaller, with a short, slender bill, while the larger Pinyon Jay has a longer bill and a square-tipped tail. Although largely blue, the Blue Jay has bold white spots in the wings and tail, a black necklace, and a pointed crest.

Mountain Bluebird

Blue with Dark Mask or Head
All three of these species are blue, with black or dark coloring on the head. The Scrub Jay has a dark mask, often set off by a white eyebrow. In the Black-throated Blue Warbler, the face and throat are black. Steller's Jay has a black head, sometimes with pale blue marks over the eyes and on the forehead, and black on the breast and upper back.

Scrub Jay

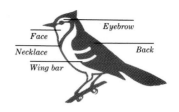

Eyebrow

Face

Necklace

Back

Wing bar

Blue Grosbeak

Indigo Bunting

Eastern Bluebird

Western Bluebird

Blue Jay

Pinyon Jay

Black-throated Blue Warbler

Steller's Jay

Black

Color and pattern differences are subtle in predominantly black birds. Check the color of the eyes and bill, and watch for iridescent or glossy plumage.

Black, Robin-sized

The slender Phainopepla and the stockier Crested Myna both have a crest and a white patch in each wing. The Phainopepla has red eyes and a dark bill; the eyes and bill of the Crested Myna are yellow. In contrast, the Rusty Blackbird lacks the crest and white wing patches, but also has yellow eyes.

Phainopepla

Black, Crow-sized or Larger

Crows and ravens are the largest all-black birds; some species, such as the Fish and the Northwestern crows, are more glossy than the others. Ravens are larger than crows and have wedge-shaped tails, while crows are generally sleeker-looking with fan-shaped tails.

Northwestern Crow

Iridescent Green or Purple

The plumage of some black birds has a green or purple iridescence in bright sunlight. The iridescent trio shown here are best told apart by size or shape: Brewer's Blackbird is robin-sized and has an iridescent purple head. The two grackles are pigeon-sized and have long, keeled tails. They overlap only in Louisiana and Texas, and their calls are different.

Brewer's Blackbird

Iridescent or Glossy

These common black birds have subtle differences: The robin-sized European Starling is iridescent green and purple with a slender yellow bill; in winter this bird is speckled and the bill dark. The Common Grackle, also robin-sized but larger than a starling, has a long, keeled tail. The sparrow-sized Brown-headed Cowbird is glossy black with a brown head.

European Starling

Head

Crested Myna

Rusty Blackbird

Fish Crow

Common Raven

Great-tailed Grackle

Boat-tailed Grackle

Common Grackle

Brown-headed Cowbird

Black-patterned

In this group, the color of the nape, wing patch, scapulars, and belly will help you identify the species. Also note bold black-and-white markings.

Colored Plumage
Several species are mainly black but have other colors in their plumage. You can recognize the Red-winged Blackbird as well as the similar Tricolored Blackbird by its red shoulders. The American Redstart has bold orange patches in the wings and tail, and a white belly. Similarly, the Bobolink is identified by its pale yellow nape, white scapulars, and white rump.

Red-winged Blackbird

Black-and-White Markings
As its name implies, the Black-and-white Warbler is a black-and-white-striped bird. The Loggerhead Shrike is also two-toned—with a dramatic black mask, gray back, and black-and-white wings and tail. The Eastern Kingbird's dark back, black crown, and black tail contrast sharply with its white underparts and white tail tip.

Black-and-white Warbler

Bold White Markings
The Black-billed Magpie is mainly black, but has a "pied" pattern—a white belly, white on the sides of its back, and white flashes in its wings. The Yellow-billed Magpie is very similar, but has a yellow bill and yellow markings on the face. The sparrow-sized Lark Bunting has bold white wing patches.

Black-billed Magpie

Black Above with Color Below
If you see a black sparrow- or robin-sized bird with a belly of a contrasting color, it is probably one of three species—the Black Phoebe with a white belly, the Rose-breasted Grosbeak with a white belly and a triangular red patch on its breast, or the Painted Redstart with a red belly. In flight, all of these birds also show white in the wings or tail.

Black Phoebe

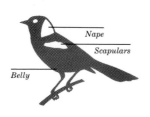

Nape
Scapulars
Belly

American Redstart

Bobolink

Loggerhead Shrike

Eastern Kingbird

Yellow-billed Magpie

Lark Bunting

Rose-breasted Grosbeak

Painted Redstart

Orange and Rufous

To distinguish orange and rufous birds, check the color of the head, face, or throat, as well as the underparts. Noting eyestripes, wing bars, a necklace, or a spotted breast will often help you identify the species.

Rufous or Chestnut Underparts
Three species have deep rufous or chestnut underparts. The ground-dwelling Rufous-sided Towhee has rufous flanks, while the Orchard Oriole, an eastern species found in shade trees, has chestnut on the underparts and breast. Familiar to all, the American Robin is known for its rufous or brick-red breast.

Rufous-sided Towhee

Black Head, Necklace, or Spots
Each of these birds has a dark field mark which contrasts with its bright orange underparts. The Northern "Baltimore" Oriole is the only orange bird with a black head and back. The Varied Thrush has a black necklace. The black spots on the breast of the Spot-breasted Oriole distinguish it from all other orange birds with a black face and throat.

Northern "Baltimore" Oriole

Black Throat and Face
Besides the Spot-breasted Oriole, three other orioles are orange with a black throat and face. The Altamira has an orange shoulder patch and one white wing bar, while the smaller Hooded Oriole has two white wing bars and lacks the shoulder patch. The Northern "Bullock's" Oriole has a black eyestripe (not face), orange eyebrow, and white wing patch.

Altamira Oriole

Head Contrasting with Body
Easily recognized by its bright orange throat, eyebrow, and crown patch, the Blackburnian Warbler also has a large white wing patch. The Black-headed Grosbeak is mainly orange-buff below, with a black head and black-and-white wings. The Western Tanager is bright yellow and black with a red face.

Blackburnian Warbler

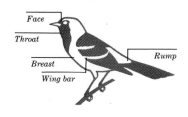

Face
Throat
Breast
Wing bar
Rump

Orchard Oriole

American Robin

Varied Thrush

Spot-breasted Oriole

Hooded Oriole

Northern "Bullock's" Oriole

Black-headed Grosbeak

Western Tanager

Yellow

The brightest yellow birds are illustrated here. Key field marks are the cap or crown, wings and tail, and markings on the face, such as spectacles or sideburns.

Predominantly Yellow

Three of our warblers are predominantly yellow, with olive-yellow or pale blue-gray upperparts. The Yellow Warbler is the brightest, and has reddish streaks on its breast. The Blue-winged has two white wing bars and a thin black line through the eye. The Prothonotary of the Southeast is a more golden color, and lacks wing bars.

Yellow Warbler

Yellow with Black on Head

Some yellow birds have conspicuous black markings or patches on the head. Wilson's Warbler has a small black cap, while the large Hooded Warbler has a black hood covering the whole head, except for a yellow face. The American Goldfinch has a black forehead and forecrown, as well as black in the wings and tail.

Wilson's Warbler

Yellow with a Bold Pattern

The Yellow-headed Blackbird is easily identified by its bright yellow head and breast. Equally distinctive, the Evening Grosbeak has bright yellow eyebrows, rump, and belly, black wings and tail, and bold white wing patches. Scott's Oriole has a black head, back, wings, and tail, with yellow on the underparts and rump.

Yellow-headed Blackbird

Yellow Breast

Against gray or olive upperparts, the yellow breast is conspicuous. The Canada Warbler has a bright yellow breast with a necklace of black spots and a white eye-ring. In addition to its yellow breast, the Kentucky Warbler has bold black sideburns, a black crown, and yellow spectacles. The Yellow-breasted Chat is told by its white spectacles and a thick, black bill.

Canada Warbler

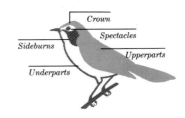

Crown
Spectacles
Sideburns
Upperparts
Underparts

Prothonotary Warbler

Blue-winged Warbler

Hooded Warbler

American Goldfinch √

Evening Grosbeak

Scott's Oriole

Kentucky Warbler

Yellow-breasted Chat

Yellow

The yellow on these birds is often combined with olive or gray upperparts. Variations in color and pattern are sometimes quite subtle. Check the head and throat, and watch for a dark mask.

Dark Mask or Markings
Yellow below and olive above, the Common Yellowthroat has a striking black mask, bordered above by white. Both the Pine and Prairie warblers also have olive upperparts and yellow underparts, but the breast is streaked with olive or black. In the Prairie Warbler, the streaks are bolder on the sides of the breast and flanks, and the face is also streaked.

Common Yellowthroat

Black on Breast
All three of these species have bright yellow in their plumage with black markings on the breast. The robin-sized Western Meadowlark has a black necklace on its breast, while the sparrow-sized Magnolia Warbler is heavily streaked with black on the breast and flanks. The Black-throated Green Warbler has a yellow face, and its breast and flanks are black.

Western Meadowlark

Gray Head
Three warblers have yellow underparts and a gray head. In the eastern Mourning Warbler and western MacGillivray's Warbler, the gray extends down to the breast as a hood; the two differ in details—a mottled black breast on the Mourning and a broken white eye-ring on MacGillivray's. The Nashville Warbler has a yellow throat and white eye-ring.

Mourning Warbler

Little Yellow
Many birds with yellow underparts and blue-gray or olive upperparts are warblers like these three. On the Northern Parula and western Orange-crowned Warbler there is enough bright yellow to consider these species "yellow birds"; but the Worm-eating Warbler begins our predominantly olive birds, with its olive upperparts, buff underparts, and black head stripes.

Northern Parula

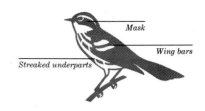

Mask

Wing bars

Streaked underparts

Pine Warbler

Prairie Warbler

Magnolia Warbler

Black-throated Green Warbler

MacGillivray's Warbler

Nashville Warbler

Orange-crowned Warbler (western)

Worm-eating Warbler

Olive

Olive birds are among the most difficult to identify by color and pattern. A brightly colored crown or eyebrow can be a good field mark. Also note the color of the eye-ring or spectacles. Watch for wing bars.

Bright Crown

The very small Ruby-crowned Kinglet as well as the sparrow-sized Green-tailed Towhee and Ovenbird all have a bright crown. The red crown patch of the Ruby-crowned Kinglet is visible only when the bird is excited. The Green-tailed Towhee has a rufous crown, with gray on the face and breast. The Ovenbird, with an orange crown, is streaked below.

Ruby-crowned Kinglet

Face Patterns

These birds have distinctive face patterns. The Red-eyed Vireo has a gray crown bordered with black, a white eyebrow, and bright red eyes. The Solitary Vireo wears a blue-gray hood and white spectacles. The very small Golden-crowned Kinglet has a white eyebrow, as well as an orange or yellow crown patch bordered with black.

Red-eyed Vireo

Yellow on Underparts

Many olive birds are yellow below. The very small Lesser Goldfinch has a black crown and a white wing patch, while the robin-sized Great Crested Flycatcher of eastern forests has a flash of rufous in the tail. In the sparrow-sized White-eyed Vireo, the yellow is confined to the flanks, and the bird has yellow spectacles and white eyes.

Lesser Goldfinch (green-backed)

Olive with Wing Bars

The sparrow-sized olive flycatchers with wing bars are very difficult to identify. In the West, a bird with yellow on the underparts and a very bold eye-ring is likely to be a Western. The Alder is best told by its call. The Ash-throated is a robin-sized bird of the Southwest; it has a gray breast, a pale yellow belly, and rufous in the tail.

Western Flycatcher

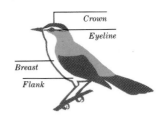

Crown
Eyeline
Breast
Flank

Green-tailed Towhee

Ovenbird

Solitary Vireo (eastern)

Golden-crowned Kinglet

Great Crested Flycatcher

White-eyed Vireo

Ash-throated Flycatcher

Alder Flycatcher

Olive or Gray

In the field, the difference between olive and gray can be subtle. On this page, you can compare the variations in hue. Good field marks for these birds are the color of the head, throat, and chin, as well as streaked or solid-colored underparts.

Olive with Wing Bars

The eastern Least Flycatcher and western Dusky Flycatcher are two more small olive flycatchers with wing bars. Both have eye-rings, and they are very hard to identify except by their calls and habitats. The two wood-pewees, the Western and Eastern, are almost identical; they are darker olive than the smaller flycatchers shown here and do not have an eye-ring.

Least Flycatcher

Olive without Wing Bars

Among sparrow-sized olive birds without wing bars, the Tennessee Warbler is easily recognized by its gray head, white eyebrow, and pale underparts. The Warbling Vireo is duller, with a thicker bill, and lacks gray on its head. The Eastern Phoebe, a flycatcher, is darker olive, but its best field marks are its upright posture and its habit of flicking its tail.

Tennessee Warbler

Gray

The Verdin and Bushtit are two very small gray birds of dry wooded and brushy country in the West. The Verdin has a yellow head, while the Bushtit is entirely dull gray. The "Slate-colored" Junco, which is the eastern form of the Dark-eyed Junco, is clear gray with a white belly and white outer tail feathers; western forms have different color patterns.

Verdin

Gray with Patterning

The Blue-gray Gnatcatcher is a tiny gray bird with a black eyebrow, white underparts, and white outer tail feathers. The eastern "Myrtle" and western "Audubon's" warblers are forms of the Yellow-rumped Warbler. Both are gray with a yellow rump, but the "Myrtle" has a white throat, while "Audubon's" has a yellow throat and a white wing patch.

Blue-gray Gnatcatcher

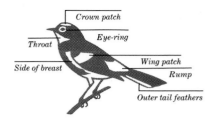

Crown patch

Eye-ring

Throat

Wing patch

Side of breast

Rump

Outer tail feathers

Dusky Flycatcher

Western Wood-Pewee

Warbling Vireo

Eastern Phoebe

Bushtit

Dark-eyed "Slate-colored" Junco

"Myrtle" Warbler

"Audubon's" Warbler

Gray

In the gray birds shown here, the color and pattern of the head—a bright cap, forehead, head stripes, or cheeks—are useful for quick identification. If you see a gray bird in flight, check the color of the tail and outer tail feathers; these clues will help you identify the species.

Gray and White
The well-named Gray Jay is largely gray, unlike other jays, but has a white head with a blackish nape. Clark's Nutcracker, a gray bird of western mountains, has bold white patches in the wings and tail. The slender Northern Mockingbird is mainly gray, but also has flashes of white in the wings and tail.

Gray Jay

Mostly Gray
Townsend's Solitaire is a slender, nondescript gray bird, with buff wing patches and white outer tail feathers. In contrast, the stocky American Dipper is entirely gray. The Gray Catbird, a slender species like its relative the Northern Mockingbird, has a black cap, and rufous coloring under the tail.

Townsend's Solitaire

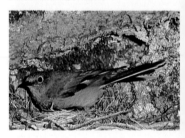

Gray Back
The Plain Titmouse of the Southwest has a gray back and paler underparts. Although the Tufted Titmouse of the East is similar, it has paler underparts, buff on the flanks, and a black forehead. The Black-capped Chickadee has a gray back, but its "chickadee" field marks are the white cheeks, and a black cap and throat.

Plain Titmouse

White Eyebrow, Black Throat
The Mountain Chickadee is similar to the Black-capped Chickadee, but has bold white eyebrows and gray flanks. Besides its white eyebrows and black throat, the Black-throated Gray Warbler has a white mustache streak and white wing bars. In the Black-throated Sparrow, the white eyebrows, white mustache streak, and a black throat are all conspicuous.

Mountain Chickadee

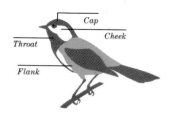

Cap
Cheek
Throat
Flank

Clark's Nutcracker

Northern Mockingbird

American Dipper

Gray Catbird

Tufted Titmouse

Black-capped Chickadee

Black-throated Gray Warbler

Black-throated Sparrow

Brown

Many of these birds have a streaked pattern on the back or underparts, and stripes on the head or crown. Look for giveaway field marks—a black spot on the breast or conspicuous white cheeks.

Crown Stripes
Both the White-throated and White-crowned sparrows have black-and-white stripes, but in the Song Sparrow, the stripes are brown and white. The White-throated has a white throat and yellow lores; the White-crowned is gray on the face, throat, and underparts. The Song Sparrow has brown-streaked underparts and a black breast spot.

White-throated Sparrow

Rufous on Crown
The Chipping and American Tree sparrows both have a solid rufous crown, but the Lark Sparrow has rufous crown stripes. In the Chipping Sparrow, the eyebrow is white bordered by a black eyeline; the American Tree Sparrow lacks these but has a dark spot on the breast. The Lark Sparrow has a breast spot, but differs in having crown stripes.

Chipping Sparrow

Brown Above with Black Throat
These three sparrow-sized brown birds have black throats. The House Sparrow is easily identified by its gray crown and conical bill. In both the Chestnut-backed and Boreal chickadees, the cheeks are white and the crown dark, but the Chestnut-backed has rich chestnut on the back and flanks, while the Boreal is duller brown.

House Sparrow

Soft Brown
These three species are soft brown. The junco can be recognized at once by its black hood. The hallmark of Say's Phoebe is the bright rufous color on its belly and undertail coverts. Unlike the other two species, the Brown Towhee is almost entirely soft brown; it also has a conical bill.

Dark-eyed "Oregon" Junco

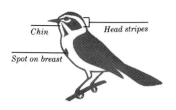

Chin

Head stripes

Spot on breast

White-crowned Sparrow

Song Sparrow

American Tree Sparrow

Lark Sparrow

Chestnut-backed Chickadee

Boreal Chickadee

Say's Phoebe

Brown Towhee

Brown

At first glance, brown birds that are streaked, spotted, or barred may look alike. To narrow your choices, note the position of these patterns. Are they on the upperparts or underparts? Also look for contrasting head patterns, such as a black mask, spectacles, or a colored eyebrow.

Streaked or Barred Above

The three wrens shown here are small brown birds with short tails. The House Wren is a dull brown, with faint barring on the wings, while the Sedge Wren is paler, buffier, and distinctly streaked. The largest of the three is the Carolina Wren, which is rufous, with a distinct white eyebrow and black bars on the wings.

House Wren

Distinct Spots or Streaks Below

The Cactus Wren, Northern Waterthrush, and Hermit Thrush are all brown birds with spots or streaks below. The Cactus Wren is the only one with a rufous cap and spots on the upperparts as well. The Northern Waterthrush is streaked below, uniform dull brown above, and has an eyebrow. The Hermit Thrush has spots only on the breast, and its tail is rufous.

Cactus Wren

Bright Rufous Above

All three of these birds are bright rufous above and feed mainly on the ground. The Brown Thrasher has a long tail and dark streaks on its underparts. The Wood Thrush and Veery are both spotted below, but in the Wood Thrush, the spots are bold and black, while in the Veery, the spots are vague and confined to the breast.

Brown Thrasher

Spectacles or Masks

These three brown birds have a distinctive face pattern. Swainson's Thrush has buff spectacles, and like other thrushes, it is spotted below. The two waxwings have a black mask, a crest, and yellow tail tips. In the Cedar Waxwing the wings are gray. The Bohemian is grayish-brown, and its wings are patterned with yellow and white.

Swainson's Thrush

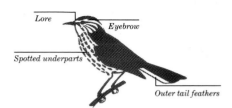

Lore

Eyebrow

Spotted underparts

Outer tail feathers

Sedge Wren

Carolina Wren

Northern Waterthrush

Hermit Thrush

Wood Thrush

Veery

Cedar Waxwing

Bohemian Waxwing

Red

A small number of birds are predominantly red or have conspicuous red underparts. In mostly red birds, note the pattern on the face: Is there an eyestripe or ear patch? In birds with red underparts and brown-streaked upperparts, watch for a red head, crown, or eyebrow. This will help you identify the species.

Rosy and Streaked
These three finches are rosy red and streaked with brown, but you can tell them apart by the amount of red on each. The western Cassin's Finch has a bright red crown and a pink breast. The Purple Finch is red on the whole head and breast, and the back is streaked with red. The smaller and browner House Finch has a red eyebrow, throat, and breast.

Cassin's Finch

Pinkish-Red
These birds are pinker than the rosy red finches. The Common Redpoll has a bright pink breast and a red crown, but its black chin makes it unlike any of the others. The Pine Grosbeak and White-winged Crossbill have similar bright pink plumage and white wing bars, but the White-winged Crossbill is smaller, and it does not have a conical bill.

Common Redpoll

Bright Red and Black
The Vermilion Flycatcher, Scarlet Tanager, and Northern Cardinal are bright red with contrasting black plumage. Right away the line through the eye and black upperparts of the Vermilion Flycatcher distinguish it from the Scarlet Tanager, which has black only on the wings and tail. The Northern Cardinal is unique with its crest and black face.

Vermilion Flycatcher

Dull to Bright Red
Unlike other red birds, the Red Crossbill is largely dull red, with dull blackish wings and tail. In contrast, the Hepatic and Summer tanagers are entirely bright red, and very similar. They differ in the color of the bill—pale yellowish in the Summer Tanager and dark in the Hepatic—and the Hepatic also has a distinctive gray ear patch.

Red Crossbill

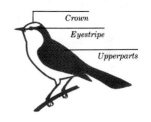

Purple Finch

House Finch

Pine Grosbeak

White-winged Crossbill

Scarlet Tanager

Northern Cardinal

Hepatic Tanager

Summer Tanager

Woodpeckers

Among the woodpeckers with patterned upperparts, important field marks to note are the location of red on the head, head stripes, and whether the upperparts are barred or checkered.

Brown or Tan with Bars

Only a few woodpeckers are brown or tan with barred upperparts. The pigeon-sized Northern "Yellow-shafted" Flicker is barred with black and brown above, while the smaller Gila and Red-bellied woodpeckers are barred with black and white on the wings and back, and tan on the rest of the body. The Gila Woodpecker has a red crown patch, and the Red-bellied is a paler tan, with a red crown and red nape.

"Yellow-shafted" Flicker

Streaked or Barred

Male Hairy and Downy woodpeckers have an almost identical pattern of checkered black and white with a small red nape patch; they differ most obviously in size—the Hairy is robin-sized, the Downy sparrow-sized. Although the Yellow-bellied Sapsucker is also intricately patterned, its markings are more muted, and it has a red crown patch and throat.

Hairy Woodpecker

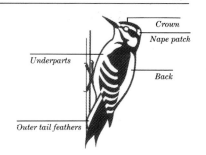

Crown

Nape patch

Underparts

Back

Outer tail feathers

Gila Woodpecker

Red-bellied Woodpecker

Downy Woodpecker

Yellow-bellied Sapsucker

Woodpeckers

To identify woodpeckers with solid backs, check the color and pattern of the head, crown, and face.

Bold Black and White

All of the woodpeckers here are more boldly patterned than the ones on the preceding pages. The first three species have common names that indicate their major field mark. Besides its brilliant red head, the Red-headed Woodpecker has a large patch of white in each wing, a white rump, and white underparts.

Black-backed Woodpecker

Bold Face Pattern

Like the woodpeckers above, these three species have black upperparts, but their face patterns make them unmistakable. The Acorn Woodpecker has a bold white face and a red cap, as well as black-and-white underparts. The large Pileated Woodpecker has a red crest, striped face, and a conspicuous white neck stripe. Lewis' Woodpecker has a red face and a gray collar and breast.

Acorn Woodpecker

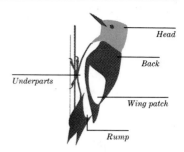

Head

Back

Underparts

Wing patch

Rump

White-headed Woodpecker

Red-headed Woodpecker

Pileated Woodpecker

Lewis' Woodpecker

How to Use the Color and Pattern Charts

By using habitat, size, behavior, and shape and posture, you have now sorted most of the hundreds of North American birds into groups, and can identify many species. You can recognize hummingbirds instantly by their very small size and hovering flight, and can quickly identify wrens by their stocky shape, short tails, and slightly downcurved bills. A forked tail has made terns easily distinguishable from gulls. The keeled tails of grackles immediately set them apart from the European Starling, with its short, square-tipped tail. For the birds that remain unidentified, color and pattern are the clincher field marks.

Color and Pattern

As we have seen, some birds are entirely one color, but most plumages are two or more colors. Depending on its species, age, and sex, every bird has colors on particular parts of its body, and the result is a pattern that is diagnostic for that bird. These distinctively colored parts of a bird can be found in more than one species, and so birders have coined a vocabulary for important body features—wing bars, crown stripes, forehead, nape, shoulder patches, and eyebrows. Whenever these features are used to identify birds in the field, they are called field marks. For example, among birds that are mainly black, only the male Bobolink has a buff nape, and so the buff nape is a good field mark for this bird. Among birds that are mainly blue, only one—the male Blue Grosbeak—has rufous wing bars; the rufous wing bars of the Blue Grosbeak are diagnostic for this species. Nearly all species have field marks that will enable you to recognize them at once. Most field marks can be seen in birds that are perching or standing, but you should also note those that can be observed only when the bird is in flight. Wing bars, wing patches, the rump, and the outer tail feathers are among the field marks that come into play when a bird is flying.

Arrangement of Color and Pattern Field Marks

To show you how field marks of color and pattern can be used to identify birds quickly and easily, I have prepared nine charts on the following pages. Only common species are included in these color charts, and the field marks listed are those that are the most conspicuous. One chart covers the ducks, a group of water birds everyone knows, and one chart covers the woodpeckers, a distinctive and easily recognized group of land birds. The seven other charts are for common songbirds. I have assigned these songbirds to basic color categories according to the color most likely to strike you first. My seven basic color charts for songbirds are blue, black or black-patterned, orange and red, yellow, olive, gray, and brown.

In each of the charts, the field marks are presented in the left column of the charts, and the species are listed across the top. (To help you in comparing birds, the species are arranged so that the most similar species are placed together.) The first field marks on the charts are the dominant color, if any. The rest of the field marks scan the bird's body from the head to the underparts, and finally to the upperparts, including the wings and tail. Some field marks, like the nape of the Bobolink or the spectacles in certain warblers and vireos, are

Color Key

■ Black or dark
□ White, grayish-white, or pale
▧ Gray or blue-gray
▧ Yellow
▧ Buff or yellowish
▧ Orange
■ Red
■ Very bright rufous
■ Brown, chestnut, or rufous
▧ Olive

▧ Green
■ Purple
■ Blue
◪ Black streaked with white
▨ Streaked
⊡ Spotted
② Two wing bars

especially helpful in distinguishing species; these "giveaway" field marks are indicated with a red dot in the left margin of the chart. At the bottom of each chart, you will find the major habitat groups for each species, as well as its size category. For each chart, if a color square is used to represent more than one color, it is indicated in the color key. For example, on the ducks chart, red stands for red or orange, and this is indicated in the key, but green only represents green, and this information does not appear in the key.

In a few cases, very similar species are not distinguished on the charts. If you are in open country, for example, and you see a streaked brown bird with bright yellow underparts and a black necklace, the chart will tell you that you have found one of the meadowlarks. To find out which one you have seen, turn to your field guide, where you will find that there are two species of meadowlarks. Comparing the subtle differences in the amount of yellow on the face and throat, the description of their voices, and the statement of the ranges of these two birds will enable you to decide which one you have seen.

Using the Charts

After you have studied the color and pattern charts, and are familiar with how they are organized, take the charts into the field and use them to identify birds. If the bird is a duck, take special note of the head: look for an overall color, or a pattern such as an ear patch, a contrastingly colored cheek or forehead. Note the color of the breast and flanks. If it is a songbird, assign it to a dominant color, or note which color seems most striking. Then refer to the appropriate color chart, and try to see as many field marks as possible. Does the bird have a crown patch? Are there eyebrows? Is the throat or breast differently colored from the rest of the body? Watch for wing bars, and take special note of any unusual feature like brightly colored shoulder patches, a spot of color on the belly, or a conspicuously colored bill. If the bird is a woodpecker, classify it according to its dominant color and pattern—brown and streaked, patterned in black and white, or mainly brown or black and without streaks. Then take a close look for any patterns or colors on the head. Since I have included only common birds in the charts, the odds are you will be able to decide what the bird is. Then confirm your identification by referring to your field guide.

Using Your Field Guide

For species not included on the charts, you can use the same method. Since you are familiar with the habitats, sizes, significant behaviors, and diagnostic shapes and postures of birds, you have a good idea of the distinctions between the bird groups we recognize in this book. In the field, you will be able to decide quickly whether the bird you see is a warbler, or a sparrow, or a heron. Note the conspicuous eyebrows, crown patches, or wing bars. More often than not, you will find that you have to choose between two or three species. Using your field notes, you can easily decide which species you have seen by checking the detailed descriptions and photographs in your guide, and then by noting the range statements.

Color and Pattern 1
Ducks

□ *White or grayish-white*
■ *Brown, chestnut, or rufous*
▨ *Buff or yellowish*
▦ *Red or orange*
Other colors: see key, p. 253
● *Giveaway field marks*

1. Northern Shoveler
2. Mallard
3. Wood Duck
4. American Wigeon
5. Green-winged Teal
6. Eurasian Wigeon
7. Cinnamon Teal
8. American Black Duck

	1	2	3	4	5	6	7	8
Dominant Color				■	▨	▦	■	■
Head	■	■	■		■	■		
Glossy	■	■	■					
● Head patch								
● Head pattern			□					
● Crown/Forehead				□				
● Ear patch					■	■		
Face patch								
● Cheeks								
Bill/Knob/Shield			■					
Collar		□						
● Neck stripe								
Underparts								
Breast	□	■	■					
● Patch at side of breast						□		
Flanks	■	▦	▨					
Upperparts: Back								
Wings								
Rear end								
Saltwater Habitats	▨	▨		▨	▨	▨	▨	▨
Freshwater Habitats	▦	▦	▦	▦	▦	▦	▦	▦
Pigeon-sized								
Crow-sized	■	■	■	■	■	■	■	■
Goose-sized								

9. Gadwall
10. Northern Pintail
11. Blue-winged Teal
12. Ruddy Duck
13. Harlequin Duck
14. Hooded Merganser
15. Bufflehead
16. Goldeneyes
17. Red-breasted Merganser
18. Common Merganser
19. Oldsquaw, winter
20. Ring-necked Duck
21. Scaups
22. Redhead
23. Canvasback
24. Common Eider
25. King Eider
26. Scoters

9	10	11	12	13	14	15	16	17	18	19	20	21	22	23	24	25	26

Color and Pattern 2
Blue

■ *Rufous brown*
Other colors: see key, p. 253

1. Painted Bunting
2. Blue Grosbeak
3. Indigo Bunting
4. Lazuli Bunting
5. Eastern Bluebird
6. Western Bluebird
7. Mountain Bluebird
8. Blue Jay

	1	2	3	4	5	6	7	8
Dominant Color		■	■	■	■	■	■	■
Head	■							
Eyebrow								
Face								□
Underparts	■						■	□
Throat/Chin					■			
Necklace								■
Breast				■	■	■		
Belly								
Upperparts: Back	■					■		
Wing bars		■		2				
Wing patch(es)								2
Forest Habitats	■	■	■	■		■		■
Open Habitats	■	■	■	■	■	■	■	
Urban and Residential Habitats	■							■
Sparrow-sized	■	■	■	■	■	■	■	
Robin-sized								■

9	10	11	12
■	■	■	■
			■
	□		
		■	
■	■		
	□	■	
	■		
			■
		□	
	■		■
		□	
■	■	■	■
		■	
■	■		■

Color and Pattern 3
Black or Black-patterned

■ *Black or dark*
□ *White or pale*
◪ *Black streaked with white*
Other colors: see key, p. 253
● *Giveaway field marks*

1. Phainopepla
2. Crested Myna
3. Rusty Blackbird
4. Crows and Ravens
5. Brewer's Blackbird
6. Great-tailed and Boat-tailed grackles
7. European Starling
8. Common Grackle

	1	2	3	4	5	6	7	8
Dominant Color	■	■	■	■	■	■	■	■
Iridescent or glossy					◪	◪	◪	◪
● **Head**								
Iridescent or glossy					■			■
Cap								
Eyebrow								
● Mask								
Eyes	■	□	□		□	□		□
Cheeks								
Bill		□					□	
● Nape								
Underparts								
Throat								
● Breast								
Belly								
Upperparts: Wings								
● Shoulder patch								
Scapulars								
Wing patch(es)	□	□						
Tail patches								
Saltwater Habitats				■		■		
Freshwater Habitats			■	■		■		
Forest Habitats	■			■	■			■
Open Habitats	■	□	■	□	□	□	■	□
Urban and Residential Habitats				■	■	■	■	■
Sparrow-sized								
Robin-sized	■	■	■		■		■	■
Pigeon-sized						■		
Crow-sized				■				

9. Brown-headed Cowbird
10. Red-winged Blackbird
11. American Redstart
12. Bobolink
13. Black-and-white Warbler
14. Blackpoll Warbler
15. Shrikes
16. Eastern Kingbird
17. Black-billed Magpie
18. Yellow-billed Magpie
19. Lark Bunting
20. Black Phoebe
21. Rose-breasted Grosbeak
22. Painted Redstart

Color and Pattern 4
Orange and Red

■ *Rufous, brown, or brick-red*
▨ *Pinkish-red*
■ *Black or dark*
▨ *Black streaked with white*
▢ *Streaked*
⊡ *Spotted*
Other colors: see key, p. 253
● *Giveaway field marks*

1. Rufous-sided Towhee
2. Orchard Oriole
3. American Robin
4. Northern "Baltimore" Oriole
5. Varied Thrush
6. Spot-breasted Oriole
7. Altamira Oriole
8. Hooded Oriole

	1	2	3	4	5	6	7	8
Dominant Color	■	■	■	■	■	▨	▨	▨
Head					■			
Crown/Forehead								
Eyebrow					▨			
● Eyestripe								
Ear patch								
Face						■	■	■
● Bill								
Underparts		■			▨	▨		
Throat/Chin						■	■	■
● Necklace					■			
Breast			■			⊡		
Belly	▢							
Flanks	■							
Upperparts								
Back								
Shoulder patch				▨		▨	▨	
Wings and Tail						■	■	■
Wing bar(s)					2		▢	2
Wing patch(es)								
Rump		■		▨				
Forest Habitats	■	■	■	■	■	■	■	■
Open Habitats		▨	▨					
Urban and Residential Habitats	■	■	■	■		■		■
Sparrow-sized		■						
Robin-sized	■		■	■	■	■	■	■

9. Northern "Bullock's" Oriole
10. Blackburnian Warbler
11. Black-headed Grosbeak
12. Western Tanager
13. Cassin's Finch
14. Purple Finch
15. House Finch
16. Redpolls
17. Pine Grosbeak

18. White-winged Crossbill
19. Vermilion Flycatcher
20. Scarlet Tanager
21. Northern Cardinal
22. Red Crossbill
23. Hepatic Tanager
24. Summer Tanager

Color and Pattern 5
Yellow

■ Gray or blue-gray
▨ Streaked
  Other colors: see key, p. 253
● *Giveaway field marks*

1. Western Tanager
2. Yellow Warbler
3. Prothonotary Warbler
4. Blue-winged Warbler
5. Wilson's Warbler
6. Hooded Warbler
7. American Goldfinch
8. Yellow-headed Blackbird

	1	2	3	4	5	6	7	8
Dominant Color	■	■	■	■	■	■	■	■
Head								
● Hood						■		
● Cap					■			
Crown/Forehead							■	
Eyebrow								
● Spectacles								
Eyeline				■				
● Mask								
● Sideburns								
Face	■				■	■		
Underparts		▨			■	■		
Throat/Chin								
● Necklace/Breast band								
Breast								
Belly								■
Flanks								
Upperparts					■	■		■
Back								
Wings and Tail	■		■	■			■	
Wing bars				②				
Wing patch								
Rump								
Freshwater Habitats		■	■			■		■
Forest Habitats	■	■		■	■	■		
Open Habitats					■		■	■
Urban and Residential Habitats		■					■	
Very Small								
Sparrow-sized	■	■	■	■	■	■	■	
Robin-sized								■

9. Evening Grosbeak
10. Scott's Oriole
11. Canada Warbler
12. Kentucky Warbler
13. Yellow-breasted Chat
14. Common Yellowthroat
15. Pine Warbler
16. Yellow-throated Vireo
17. Prairie Warbler

18. Meadowlarks
19. Magnolia Warbler
20. Townsend's Warbler
21. Hermit Warbler
22. Black-throated Green Warbler
23. Mourning and
    MacGillivray's warblers
24. Nashville Warbler
25. Northern Parula
26. Orange-crowned Warbler,
    western

9	10	11	12	13	14	15	16	17	18	19	20	21	22	23	24	25	26

Color and Pattern 6
Olive

■ Olive or olive-brown
□ White or pale
■ Brown, chestnut, or rufous
▨ Streaked
Other colors: see key, p. 253
● *Giveaway field marks*

1. Orange-crowned Warbler, western
2. Worm-eating Warbler
3. Ruby-crowned Kinglet
4. Green-tailed Towhee
5. Ovenbird
6. Red-eyed Vireo
7. Solitary Vireo, eastern
8. Golden-crowned Kinglet

	1	2	3	4	5	6	7	8
Dominant Color	■	■	■	■	■	■	■	■
Head				■			■	
Crown patch			■					
Crown/Forehead				■	■			■
● Crown stripes		■						
Eyebrow	□					□		
● Spectacles							□	
Eye-ring								
● Eye						■		
Facial stripes								■
Underparts	■	□			▨	□	□	
Throat/Chin								
Breast					■			
Belly								
Flanks							■	
Upperparts: Wing bars			2				2	2
Wing patch								
● Tail								
Freshwater Habitats								
Forest Habitats	■	■	■	■	■	■	■	■
Open Habitats				■		■		
Urban and Residential Habitats			■	■				■
Very Small			■					■
Sparrow-sized	■	■		■	■	■	■	
Robin-sized								

9. Lesser Goldfinch, green-backed
10. Great Crested and
    Ash-throated flycatchers
11. White-eyed Vireo
12. Western Flycatcher
13. Alder, Least, and
    Dusky flycatchers
14. Wood-pewees
15. Tennessee Warbler
16. Warbling Vireo
17. Eastern Phoebe

Color and Pattern 7
Gray

■ *Buff*
☑ *Streaked*
Other colors: see key, p. 253
● *Giveaway field marks*

1. Verdin
2. Bushtit
3. Dark-eyed "Slate-colored" Junco
4. Blue-gray Gnatcatcher
5. Yellow-rumped "Myrtle" Warbler
6. Yellow-rumped "Audubon's" Warbler
7. Gray Jay
8. Shrikes

	1	2	3	4	5	6	7	8
Dominant Color	■	■	■	■	▨	▨	■	■
● **Head**	buff							
Head stripes								
Crown/Cap								
Forehead								
Eyebrow				■				
Eye-ring								
● **Mask**								■
Cheeks								
Face							□	
● **Nape**							■	
Underparts				□	▨	▨	□	
Throat/Chin					□	buff		
● **Sides of breast**					buff	buff		
Belly			□					
Flanks								
Upperparts: Wings								■
Wing patch(es)								□
Rump					buff	buff		
Tail				■				■
Outer tail feathers			□	□				□
Saltwater Habitats					buff			
Freshwater Habitats								
Forest Habitats	■	■	■	■	■	■	■	■
Open Habitats	buff		buff					buff
Urban and Residential Habitats			■	■				
Very Small	■	■		■				
Sparrow-sized			■		■	■		
Robin-sized							■	■
Pigeon-sized								

9. Clark's Nutcracker
10. Northern Mockingbird
11. Townsend's Solitaire
12. American Dipper
13. Gray Catbird
14. Plain Titmouse
15. Tufted Titmouse
16. Black-capped and
    Carolina chickadees
17. Mountain Chickadee

18. Black-throated Gray Warbler
19. Black-throated Sparrow
20. Pyrrhuloxia

9	10	11	12	13	14	15	16	17	18	19	20

Color and Pattern 8
Brown

■ *Brown, rufous, or chestnut*
■ *Very bright rufous*
□ *White, grayish-white, or pale*
■ *Black or dark*
▨ *Buff*
▨ *Streaked*
⊡ *Spotted*
Other colors: see key, p. 253

1. White-throated Sparrow
2. White-crowned Sparrow
3. Song Sparrow
4. Fox Sparrow, eastern
5. Chipping Sparrow
6. American Tree Sparrow
7. Lark Sparrow
8. House Sparrow

● *Giveaway field marks*

	1	2	3	4	5	6	7	8
Dominant Color	■	■	■	▨	■	■	■	■
● **Head:** Hood								
Crown/Cap					■	■		▨
Head stripes	▨	▨	▨	▨			▨	
Eyebrow					□	▨		
Eyestripe					■	■		
● Mask								
● Spectacles								
● Cheeks								
Lores	▨							
Nape								■
Underparts		▨	▨	▨	▨	▨	□	▨
Throat/Chin	□							■
Breast								
● Spot on breast			■			■	■	
Belly								
Flanks								
Upperparts: Back								
Wing bars						②		
Tail								
Tail tip								
Outer tail feathers							□	
Freshwater Habitats								
Forest Habitats	■	■	■	■	■			
Open Habitats	▨	▨	▨	▨	▨	▨	▨	▨
Urban and Residential Habitats	■		■		■		■	■
Very Small								
Sparrow-sized	■	■	■	■	■	■	■	■
Robin-sized								

9. Chestnut-backed Chickadee
10. Boreal Chickadee
11. Dark-eyed "Oregon" Junco
12. Say's Phoebe
13. Wrentit
14. Brown Towhee
15. House Wren
16. Sedge Wren
17. Carolina Wren
18. Winter Wren
19. Cactus Wren
20. Waterthrushes
21. Hermit Thrush
22. Brown Thrasher
23. Wood Thrush
24. Veery
25. Swainson's Thrush
26. Waxwings

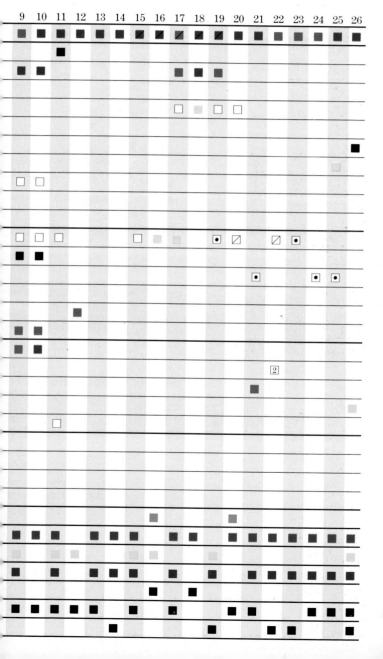

Color and Pattern 9
Woodpeckers

- ◪ *Brown or tan*
- ◪ *Black streaked with white*
- ▧ *Streaked*
- ⊡ *Spotted*
  Other colors: see key, p. 253
- ● *Giveaway field marks*

1. Northern "Yellow-shafted" Flicker
2. Northern "Red-shafted" and "Gilded" flickers
3. Gila Woodpecker
4. Golden-fronted Woodpecker
5. Red-bellied Woodpecker
6. Hairy Woodpecker
7. Downy Woodpecker
8. Yellow-bellied Sapsucker

	1	2	3	4	5	6	7	8
Dominant Color	◪	◪	◪	◪	◪	◪	◪	◪
● **Head**								
Head stripes						◪	◪	◪
● Crest								
Crown patch		■	■	■				
Crown/Cap	■				■			■
Face			■	■	■			
Cheeks		■						
● Mustache	■	■						
Nape	■				■			
Nape patch						■	■	
Underparts	⊡	⊡	■	■	■	☐	☐	☐
Throat/Chin								■
Collar								
Breast								
Flanks								
Upperparts: Back						☐	☐	
Wing patch(es)								☐
Rump	☐	☐						
Forest Habitats	■	■	■	■	■	■	■	■
Open Habitats	■		■					
Urban and Residential Habitats		■	■	■	■	■	■	■
Sparrow-sized							■	
Robin-sized			■	■	■	■		■
Pigeon-sized	■	■						
Crow-sized								

9. Red-naped Sapsucker
10. Ladder-backed Woodpecker
11. Red-cockaded Woodpecker
12. Three-toed Woodpecker
13. Black-backed Woodpecker
14. White-headed Woodpecker
15. Red-headed Woodpecker
16. Red-breasted Sapsucker
17. Acorn Woodpecker

18. Pileated Woodpecker
19. Williamson's Sapsucker
20. Lewis' Woodpecker
21. Strickland's Woodpecker

Voice

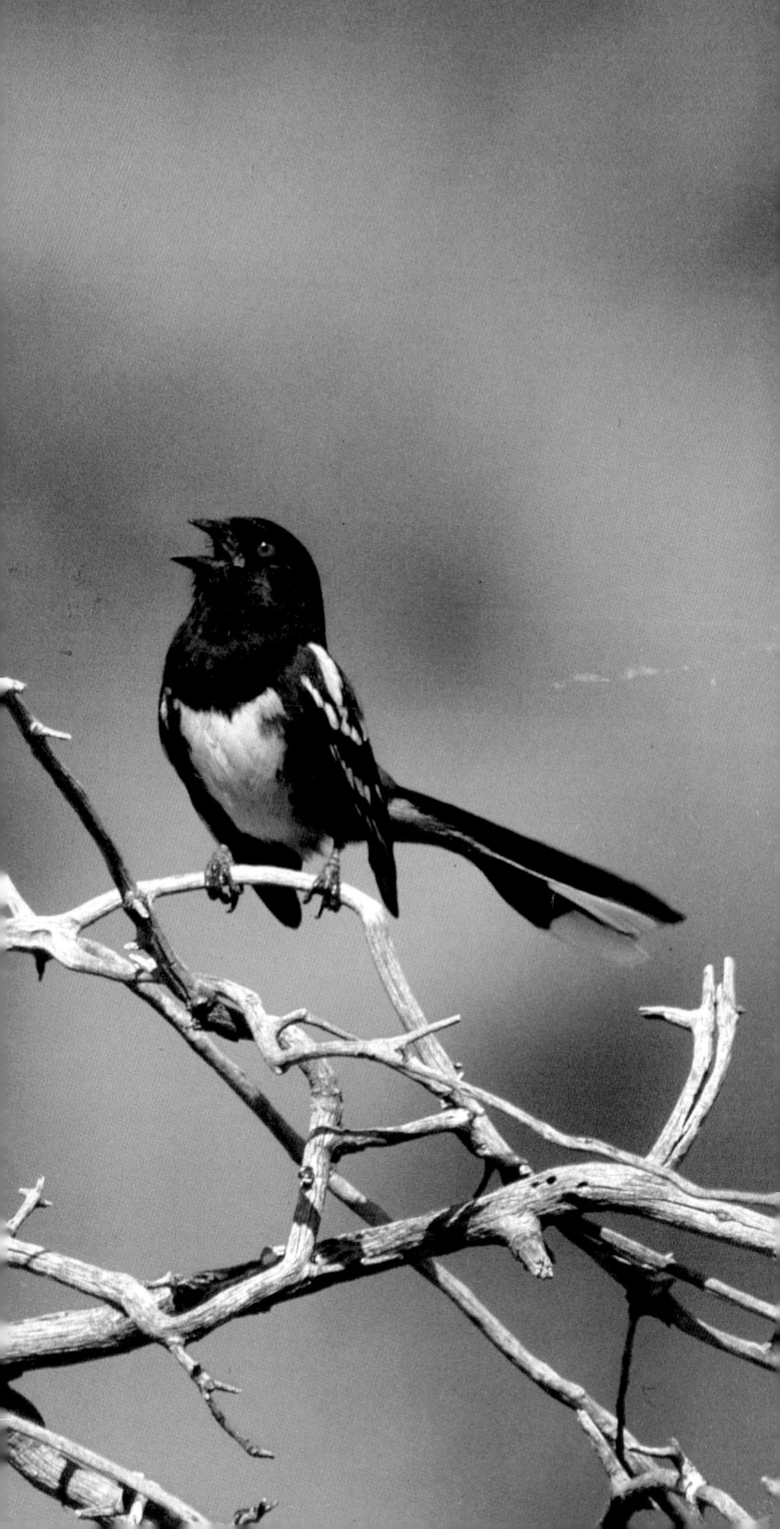

Voice

Most North American birds have distinctive voices, and to a trained ear, the calls and songs of birds offer a rich array of vocal field marks, some even more diagnostic than color and pattern. In fact, an experienced birder who is familiar with individual bird calls can identify a bird on a field trip by its voice alone. As a beginner, however, it is easier for you to learn to identify a bird by sight first and then listen to its call to reinforce your identification. You should master the field marks we have already discussed—habitat, size, behavior, shape and posture, and color and pattern—before you start learning some of the interesting and beautiful sounds of birds.

Calls and Songs

Nearly all bird sounds can be classified as either calls or songs. Calls are generally short and simple, often of one syllable. They serve many purposes—to express alarm, maintain contact between members of a flock, threaten other birds, or scold predators. In contrast, songs are usually longer and more complex than calls; they are used mainly by males to advertise their claim to a nesting or feeding territory or to attract a mate. Calls can be heard at any time of year, while songs are most often heard in spring and early summer when birds are nesting. Songs are also heard occasionally during the fall, and from birds that maintain feeding territories in winter.

You might think that all birds that sing songs are songbirds, but this is not true. Ornithologists use this term to describe a large group of land birds with an intricate vocal apparatus, but songbirds are not the only species capable of songs. In fact, many other birds have songs and all except a very few birds have calls. Because it is not always obvious whether a bird sound is a call or song, this distinction will not immediately help you identify birds. Instead, in this chapter I am concentrating on the sound itself, its quality or tone and its pattern, rather than on whether it is a call or song. In previous chapters I focused on what you can see; here we are focusing on what you can hear.

Using Voice Alone

Sometimes recognizing a bird's voice is essential for identification because other field marks are insufficient. At night, owls and nightjars are best told by their calls, and in fact, these birds seem to recognize one another in the dark by voice rather than by sight. Similarly, a few birds, such as some flycatchers, are almost identical in color and pattern, but are easily distinguished by their different voices. The birds themselves seem to use vocal clues to recognize members of their own species; their calls serve the same purpose as the distinctive colors and patterns of other birds.

Transliterations and Descriptions of Bird Voices

Because voice is such a valuable and interesting field mark, ornithologists and birders have developed several methods for describing the many sounds that birds make. The oldest methods for conveying bird voices are verbal transliterations and descriptions. In a transliteration, the sound made by a bird is rendered as closely as possible in words. Although each of us has a different ear for bird

The calls and songs of birds
described on the following pages
are divided into three groups—
simple sounds, patterns of notes,
and varied or repeated phrases.

calls, all of us would agree that the words *chick-a-dee* are a good
approximation of the standard call of most species of chickadees, and
that the Whip-poor-will does indeed say *whip-poor-will*. But it may
not help you very much to read that the Magnolia Warbler sings a
hurried *wee-o wee-o wee-chy*, or that the song of a Common
Yellowthroat sounds like *ter-weechy ter-weechy ter-weechy*. You
must first hear these songs to understand their transliterations.
Another way the sounds of birds are conveyed is through description.
For example, a Canada Goose "honks," or the House Sparrow "chirps
and cheeps repeatedly." Birders use many other terms, such as lisp,
trill, warble, cooing, and hooting; they speak of phrases and patterns,
and of repeated phrases and flutelike notes.

Recordings
The voices of most North American birds have been recorded, and
commercial records and cassettes are readily available. But cassettes
or records are not a very convenient way to learn to recognize a bird
call the first time you hear it. They are valuable tools when you
already know many bird songs and are interested in a particular
species or want to confirm what you have heard in the field.

Mastering Bird Voices
The easiest way to learn bird calls and songs is to identify a bird using
visual field marks, and then to note any sound the bird makes. Begin
in your backyard or urban neighborhood, where you have already
identified birds and are familiar with them. Listen to their songs and
calls several times. Once you have learned to recognize the voices of
these birds, you can use them for comparison when you go farther
afield.
As a next step, you should try to write down what you have heard.
Use simple descriptions or transliterations. Compare these notes with
the descriptions in the field guide. There you will find that I have used
either description or transliteration, whichever method seemed more
appropriate. (If the field guide does not include a description of the
voice, this is because a species does not have a call or song that is
helpful in identification.) In going through this exercise you will not
only become a better listener, but will also learn how to note bird
voices in a way that makes them useful field marks.

Using the Illustrations
So that you can become familiar with some of the descriptive terms
used in my field guides, in the following pages I am illustrating three
groups of common birds with descriptions of twelve types of calls and
songs. The text explains the quality or pattern the voices of these
birds share. (Sounds that need no description, like the cawing of
crows or the quacking of ducks, are not included.) Once you can
identify birds by their voices alone, you are well on the way to
becoming an expert birder.

Simple Sounds

Lisp
A lisp is a very high-pitched, sibilant note. Its form varies, depending on the species. The Brown Creeper gives a single lisp, while the Golden-crowned Kinglet has two lisping notes. The Cedar Waxwing usually repeats the lisp several times; a flock of waxwings creates a continuous lisping sound. All of these notes are calls.

Brown Creeper

Trill
A trill is a simple series of short, similar notes delivered very rapidly. A trill is often delivered on one pitch, but the brief trill of the Orange-crowned Warbler drops abruptly in pitch at the end. The song of the Dark-eyed Junco can sound dry and mechanical, but may also be musical.

Chipping Sparrow

Insectlike Buzz
Buzzy, insectlike songs are most often heard in sparrows that inhabit grasslands, although a few warblers, like the Blue-winged and Golden-winged, also have songs that belong in this category. These dry, buzzy songs seem to carry farther in grasslands than the lower-pitched songs typical of forest birds.

Clay-colored Sparrow

Cooing
Most often associated with pigeons and doves, cooing is a low-pitched song or call. It is also given by a few other species such as the Greater Roadrunner. In some species, cooing has a mournful or hollow quality, and it is usually given as a series of notes, sometimes in different pitches.

Greater Roadrunner

The bird calls and songs described here are usually simple—either a single note, a trill, a buzz, or a series of low-pitched cooing notes.

Golden-crowned Kinglet

Cedar Waxwing

Orange-crowned Warbler

Dark-eyed "Slate-colored" Junco

Savannah Sparrow

Grasshopper Sparrow

Rock Dove

Mourning Dove

Patterns of Notes

Clear Whistle
Most birds with clear, whistled songs have a distinctive pattern, whether it is the plaintive series of notes of a Field Sparrow, the sweetly whistled phrases of a White-throated Sparrow, or the breezy, far-carrying song of an Eastern Meadowlark.

White-throated Sparrow

Warble
A warble is a melodic song, often with a slight burr, that is delivered so rapidly that it cannot be rendered clearly in words as a transliteration. The songs of the three species pictured here are patterned, and it is their canarylike quality that makes them good examples of warbles.

Purple Finch

High-pitched, Patterned Songs
The songs of many warblers are very high-pitched, with a constant pattern that is quite easy to transliterate. (Despite their name, the songs of warblers are not warbles.) A common song of the Yellow Warbler, for example, can be rendered as *sweet-sweet-sweet-sue-so-sweet*. These patterns are easy to learn, once you are familiar with the species.

Yellow Warbler

Songs without Clear-cut Phrases
The songs of these species have a rapid, almost ecstatic quality that makes them easy to single out. Some songs in this category are musical enough to be considered warbles, but those of the birds shown here are higher-pitched, and those of the two wrens have a chattering quality.

Horned Lark

These four types of bird sounds all display a distinct pattern of notes, which may be rapid, as in some of the wrens, or slower and more melodic, as in a warble or the clear whistled song of a White-throated Sparrow.

Field Sparrow

Eastern Meadowlark

House Finch

Fox Sparrow (western)

Black-throated Green Warbler

Blackburnian Warbler

Marsh Wren

House Wren

Varied or Repeated Phrases

Flutelike Notes

The patterned songs of the Western Meadowlark and the two thrushes have a harmonic sound, which has earned them the adjective flutelike. It is the melodious quality of the notes, rather than the patterns, that makes the songs in this category distinctive.

Western Meadowlark

Repeated Phrase or Note

Many species, including the three shown here, have songs or calls that consist of the same note or phrase repeated several times in succession. For example, in Steller's Jay, a frequently heard call sounds like *shook-shook-shook*, and in the Common Yellowthroat and Northern Cardinal, the song consists of such repetitions.

Common Yellowthroat

Varied Phrases

A number of songbirds, including the American Robin and the three species shown here, have songs that are made up of phrases, but the phrases are varied, and not repeated immediately. It is the pattern of notes rather than the quality or tone of the sound that defines this category.

Summer Tanager

Varied, Repeated Phrases

The songs of these species are made up of different phrases that vary in quality or tone. They may be repeated once, as in the Indigo Bunting, or several times, as in the Northern Mockingbird, before a new phrase is chosen and repeated in its turn. Many of these birds sing tirelessly all day long.

Indigo Bunting

Here are some of the more
complex bird songs, involving a
variety of notes, varied or
repeated phrases, and even the
interweaving of repeated series of
phrases.

Swainson's Thrush

Hermit Thrush

Steller's Jay

Northern Cardinal

Rose-breasted Grosbeak

Western Tanager

Northern Mockingbird

Curve-billed Thrasher

A Final Word

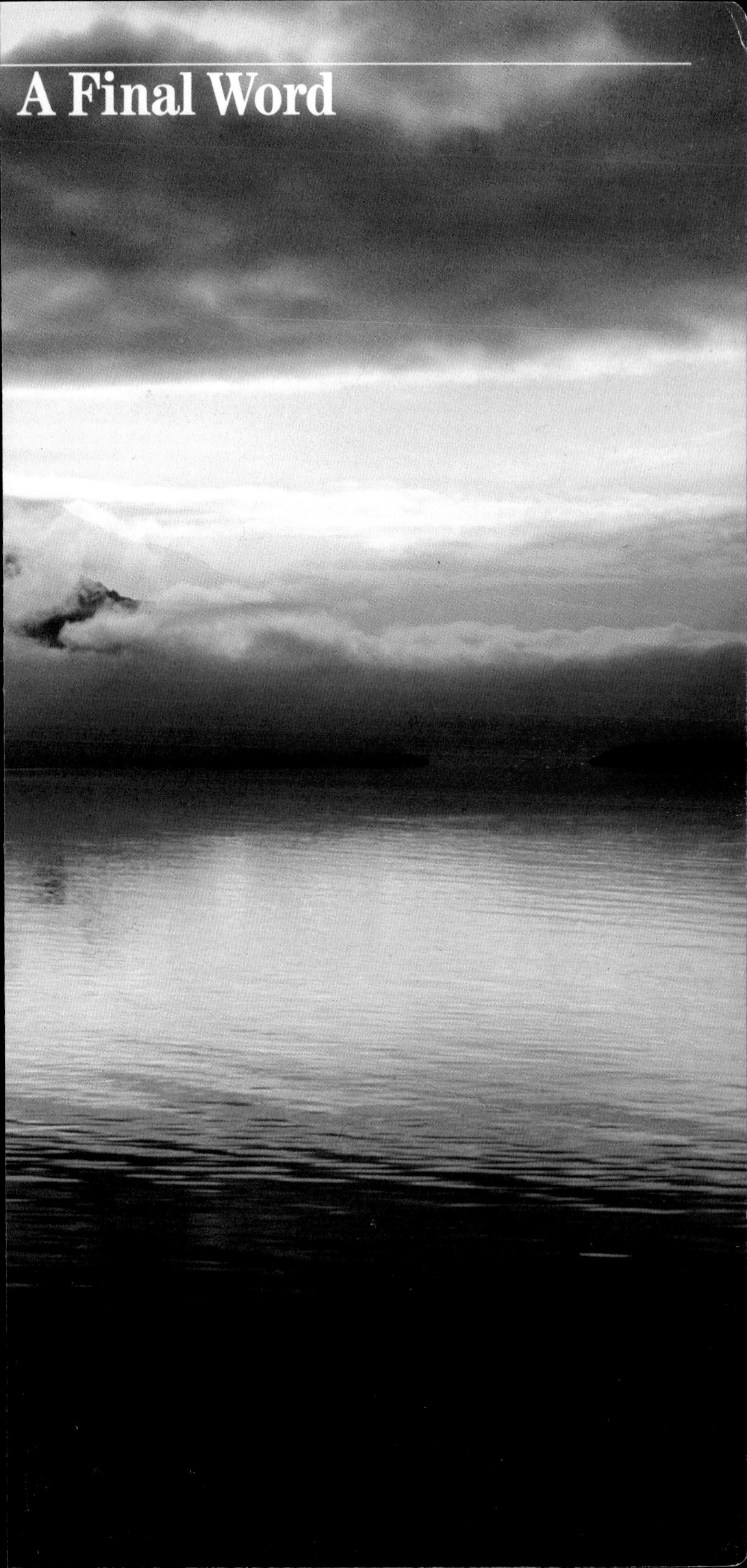

A Final Word

Now that you have studied this handbook carefully and are familiar with the key features of birds—their field marks—you are well on the way to mastering the techniques experts use to identify birds. As you embark on this rewarding lifelong pursuit, some general hints may be helpful.

Notebook

I have found that a notebook is essential for studying birds and learning their field marks. Since none of us has a perfect memory, detailed notes are the only means of making sure that your on-the-spot impressions of a bird are safely on record for use later in identifying the species. Taking notes is also a good way to sharpen your powers of observation, because when you are deliberately looking for things to write down you are bound to notice more than you would in a casual glance at a bird.

In the field, I use a pocket-sized notebook with a stiff cover so that the pages lie flat when I am taking notes and do not become dog-eared. I have been using the same brand for fifteen years, one whose covers come in a variety of colors. I buy several at a time, and use one color for birding in my own area, and different colors for taking notes on extended trips away from home. Using different colors enables me to recognize my current notebook for local observations, or quickly find the one I used on a trip to California or Colorado. It is best to write on only one side of each page; when I have filled a page, I turn to a new one and keep the used pages out of the way with a paper clip. I prefer to use a pen because it is permanent, but pencils have the advantage of not freezing up in very cold weather. Short, stubby pencils are better than long, sharp-pointed ones, because you can carry several, and the lead is not as likely to break.

Whenever you are out birding, you should write down the date, place, weather conditions, and, if you are at the shore, whether the tide is in or out. Then as soon as you see a bird, note field marks in the order presented in this handbook. First, write down the habitat you are in. Then record the bird's size. Try to describe at least one observation on the bird's behavior. Next, write down whatever strikes you about its shape and posture, and finally note the bird's most conspicuous features of color and pattern. If you are aware of a call or song, try to describe it or approximate it in words. After you have noted field marks a few times, you will find that you are checking these categories easily and quickly. While you will always need habitat and size, you may only have to jot down a few field marks in other categories to identify a species. As you gain experience, this process will become more and more automatic.

To help you get started with your notebook, a sample notebook page with entries for you to fill in is shown on page 288. I suggest that you set up the pages of your notebook in the same way. This will help you remember the features and field marks to observe when you see a bird.

If you start taking field notes when you first begin birding, you are likely to continue to do so after you have mastered the techniques of rapid identification and are learning about the natural history of birds

How much it enhances the
wildness and the richness of the
forest to see in it some beautiful
bird which you never detected
before!

Henry D. Thoreau, 1853
Journal

firsthand. Never hesitate to take notes on something just because you think it is already well known to other birders or to ornithologists. It may not be, for birding is a subject in which amateurs can still make important contributions. The important thing is not what is known to science, but what is known to you. There is a vast difference between reading a fact in a book and finding it out for yourself.

I remember an afternoon I spent walking along a pebbly beach in Massachusetts with an experienced birder. Things were pretty quiet; the only birds we could find were some Ring-billed Gulls, a common species whose natural history is well documented in books and technical papers. I began taking notes on the calls these gulls were giving as they fed. My companion saw that I had my notebook out and asked: "What are you doing that for? The calls of the Ring-billed Gull are in every field guide." I was only taking those notes to pass the time, but because of what I wrote down that afternoon, I can now distinguish Ring-billed and Herring gulls by their calls alone, as they fly high above the streets of New York City. Field guides will correctly tell you that the calls of the Ring-billed Gull are higher-pitched than those of the Herring Gull, but only when you have carefully noted this difference yourself can you use it for identification.

As your notes accumulate, their usefulness increases. They are an irreplaceable record of your field trips, and of everything you have learned about birds, their habits, and their ecology. A set of carefully recorded notes will become the cornerstone of your birding library, perhaps more valuable than any bird book you can buy.

Binoculars

Like a field notebook, a pair of binoculars is an important part of a birder's equipment. "Opera glasses" do not make good birding binoculars, but a pair of binoculars that is good for birding will do very well in an opera house. When you go shopping for binoculars, look for a pair with certain specifications: The best kinds for birding are 7×35 or 8×40 binoculars, but if you are steady-handed, you may be able to use a pair of 10×50s. The first number indicates the magnification; a pair of 7×35s will magnify an image seven times, while a pair of 10×50 binoculars magnifies a bird ten times. But a high magnification like $10 \times$ is difficult to hold steady, because it also magnifies your hand movements ten times. The second number is the diameter of the outer lens in millimeters; it indicates how much light will enter the binoculars. You should avoid binoculars in which the second number is less than five times the magnification, because the outer lens will then be too small to admit enough light when you are in a shady forest. A measurement of more than five times the magnification will let in too much light, and will be difficult to use if you are birding in very bright sunlight—on a beach, in the desert, or in a snowy field.

Your binoculars should be lightweight but sturdy and waterproof. Coated lenses cut down the glare and reduce distortion. Binoculars with center focusing are easier to use than a pair in which each eyepiece has to be focused separately. There are binoculars for every pocketbook; you should buy the best pair you can afford and then take proper care of them. Store them in their case when you are not using them, and keep them clean with lens paper and lens cleaning fluid. Never wipe the lenses with coarse paper or cloth, because this may scratch the coating. Be especially careful about sand and dried salt spray if you have been birding at the shore.

Using binoculars in the field takes a little practice. To find a bird with your glasses, look at it with the naked eye, and then, without changing the direction of your gaze, quickly bring the binoculars up between your eyes and the bird. You will usually find the bird in view in your binoculars. If you have difficulty doing this at first, try following a branch outward to where you have seen the bird or try using a nearby object as a point of reference. In a very short time, you will have no trouble fixing your binoculars on a bird quickly.

Telescopes

Once you are "hooked," and have become a serious birder, you will probably want to buy a telescope for checking shorebirds on a distant sandbar or scanning a flock of gulls or ducks resting on the water. Like binoculars, a telescope should be lightweight and have coated lenses. The magnification should vary from $20 \times$ to $60 \times$, and the telescope may have fixed lenses or a zoom lens, whichever you find more comfortable. It is a good idea to use the telescopes of birding companions, and then buy the kind that best suits you.

In order to use a telescope, you will need a tripod. While your telescope should be lightweight, the tripod should be very sturdy, so that the telescope doesn't vibrate in a strong wind. For viewing birds in flight, you may want to mount your telescope on a gunstock; clamps are available that can be used to attach a telescope to the window of your car, a blessing on a cold and rainy day along the shore.

Attracting Birds

Birds may be shy, but they are also curious. You can often attract small birds and even hawks or owls by making squeaking noises that sound like a bird in distress. I have used a simple squeaking sound for years, and have not only brought timid sparrows and warblers out of thick underbrush, but have unwittingly attracted Common Barn-Owls, Short-eared Owls, Cooper's Hawks, and even foxes, weasels, and house cats, all looking for a possible meal.

Clothing

Your field clothes should be warm, waterproof, and comfortable. It is best not to wear bright colors that will attract the attention of birds, and fabrics should not rustle or get snagged on thorns. Footgear is important. It should be sturdy and thorn-proof in deserts and rocky areas, and waterproof or easy to dry in wet places. I prefer running shoes or sneakers in most habitats, but wear good boots in deserts where cactus spines or poisonous snakes are a potential hazard. Some birders can ignore insect bites, but if you are like me, you will want to carry insect repellent.

Field Guides as Tools

Once you have determined the field marks of a bird you have seen,

And as the hermit's evening hymn goes up from the deep solitude below me, I experience that serene exaltation of sentiment of which music, literature, and religion are but the faint types and symbols.

John Burroughs, 1872
Wake-Robin

I heard it here.

Theodore Roosevelt *(Trying to relocate a Blue-gray Gnatcatcher for John Burroughs.) Undated.*

you need a field guide to confirm its identification. Simply look up the species name in the index and then turn to the illustration and text to check your identification. While there are several excellent field guides available, you may wish to use my field guides that are coordinated with this handbook, namely, *An Audubon Handbook: Eastern Birds* and *An Audubon Handbook: Western Birds.*

In my companion guides, the field marks you have studied here are clearly described and illustrated. For example, the first feature to notice—habitat—appears prominently at the top of the page, along with a special size "yardstick," representing one of the seven size groups you have learned. The other field marks, such as behavior or shape and posture, are noted in the text accounts. Moreover, the arrangement of the text-and-picture descriptions by similarities, such as habitat, lookalikes, or related species, allows you to look up a species quickly and to compare it with other species. For example, the ducks are grouped with other similar water birds, like geese, swans, grebes, and loons. Another convenience in the companion field guides is that the illustrations and text description for a species are on a single page. For each bird, there are from one to four color photographs, including close-ups showing different plumages and postures, as well as views of birds in flight. Because swans, geese, ducks, raptors, and gulls are often identified on the wing, special color comparisons show these groups in flight. Finally, using the unique charts of field marks in this handbook, you can easily find a species or a group of species in one of the field guides.

Mastering the Art of Birding

If you can't identify a bird the first time you see it, don't worry. You will certainly find it again, and you will probably see enough the second time to find it in your field guide. Eventually you will know certain species quite well and you can use them for comparison with unfamiliar birds. The chances are that you will soon know whether it is a warbler, or a plover, or a gull. The more species you get to know, the easier it is to learn new ones.

Finally you will reach the point where you know all the common species in your area, and want to try other localities. The following section on page 290, Guides to Birding Localities, will help you find birding sites, both near home and on trips to distant parts of the country. Then it is the rarities—the species birders are willing to drive miles to see—that will attract your attention. Rarities, species that turn up far from their normal range, are what add spice to bird identification after you have become an expert. When a rarity is found in an area, word spreads quickly among local birders. In many places there are special telephone numbers with information on rare birds and directions for locating them. To find out whether there is a "Rare Bird Alert" telephone number in your area, contact your local bird club, your library, local museum, or a nature center.

A Final Word
Notebook

This notebook page is specially designed as a model for your own notebook. If you write down field marks in the order presented here, you will quickly master the art of field indentification of birds.

Species

Date	*april 12th*
Place	*Backyard at feeder*
Weather	*Sunny day*
Tide	

Habitat: Saltwater

Freshwater

Forest

Open

Urban and Residential ✓

Size: Very Small

Sparrow-sized ✓

Robin-sized

Pigeon-sized

Crow-sized

Goose-sized

Very Large

Behavior: On or over water

On ground, in bushes, and in trees *Eats seeds at feeder*

In flight *Seems to bounce through air*

Shape and Posture

Slender   Stocky   Horizontal   Vertical

Shape: Head and Neck

Bill   *Short, conical*

Wings

Tail or Legs

Color and Pattern: Dominant color   *Brown-streaked*

Head   *Red on eyebrow, chin and throat*

Underparts   *Brownish-streaked*

Upperparts

Voice: Simple sounds

Patterns of notes   *Cheerful warble*

Varied or repeated phrases

Other

Guides to Birding Localities
Arizona to Missouri

Arizona
William A. Davis and Stephen M. Russell
Birds in Southeastern Arizona
Tucson Audubon Society, revised edition, 1984

James A. Lane and Harold R. Holt
A Birder's Guide to Southeastern Arizona
L & P Press, Denver, fifth edition, 1986

California
James A. Lane and Harold R. Holt
A Birder's Guide to Southern California
L & P Press, Denver, revised edition, 1985

Jean Richmond
Birding Northern California
Mount Diablo Audubon Society, Walnut Creek, California, 1985

Colorado
Harold R. Holt and James A. Lane
A Birder's Guide to Colorado
L & P Press, Denver, 1987

Connecticut
Noble S. Proctor
25 Birding Areas in Connecticut
Pequot Press, Chester, Connecticut, 1978

Delaware
John J. Harding and Justin J. Harding
Birding the Delaware Valley Region
Temple University Press, Philadelphia, 1980

Claudia Wilds
Finding Birds in the National Capital Area
Smithsonian Institution Press, Washington, D.C., 1983

District of Columbia
Claudia Wilds
Finding Birds in the National Capital Area
Smithsonian Institution Press, Washington, D.C., 1983

Florida
James A. Lane and Harold R. Holt
A Birder's Guide to Florida
L & P Press, Denver, revised edition, 1984

Georgia
Joe Greenberg and Carole Anderson
A Birder's Guide to Georgia
Georgia Ornithological Society, Cartersville, second edition, 1984

Illinois
Elton Fawks, edited by Paul H. Lobik
Bird Finding in Illinois
Illinois Audubon Society, Downers Grove, 1975

For many states there are specialized books that list the best birding localities and provide directions on how to reach each site, information on the species you can expect to see, and often discussions of habitats and ecology. The following is a state-by-state list of some of the most useful guides to finding birds. They are available at bookstores, nature centers, and public libraries. Note that some states do not have comprehensive local guides.

Indiana
Charles K. Keller, Shirley A. Keller, and Timothy C. Keller
Indiana Birds and Their Haunts: A Checklist and Finding Guide
Indiana University Press, Bloomington, second edition, 1986

Iowa
Peter C. Petersen, Jr., editor
Birding Areas of Iowa
Petersen Book Company, Davenport, Iowa, revised edition, 1986

Kansas
John L. Zimmerman and Sebastian Patti
A Bird Finding Guide to Kansas and Western Missouri
University Press of Kansas, Lawrence, Kansas, 1987

Louisiana
James Whelan
A Bird Finding Guide to the New Orleans Area
Louisiana Nature Center, New Orleans

Maine
Elizabeth Cary Pierson and Jan Erik Pierson
A Birder's Guide to the Coast of Maine
Down East Books, Camden, Maine, 1981

Maryland
Claudia Wilds
Finding Birds in the National Capital Area
Smithsonian Institution Press, Washington, D.C., 1983

Massachusetts
Edith F. Andrews
Birding Nantucket
Privately published, 1984

Michigan
John Eastman, editor
Enjoying Birds in Michigan
Michigan Audubon Society, Kalamazoo, fourth edition, 1987

Minnesota
Kim Eckert
A Birder's Guide to Minnesota
Minnesota Ornithologists' Union, Minneapolis, revised second ed., 1983

Missouri
Richard A. Anderson and Paul E. Bauer
Birds of the St. Louis Area
Webster Groves Nature Society, Webster Groves, Missouri, 1968

John L. Zimmerman and Sebastian Patti
A Bird Finding Guide to Kansas and Western Missouri
University Press of Kansas, Lawrence, Kansas, 1987

Guides to Birding Localities
New Jersey to Wisconsin

New Jersey
William J. Boyle
A Bird Finding Guide to New Jersey
Rutgers University Press, New Brunswick, New Jersey, 1986

John J. Harding and Justin J. Harding
Birding the Delaware Valley Region
Temple University Press, Philadelphia, 1980

New Mexico
Dustin Huntington and Dale A. Zimmerman, editors
New Mexico Bird Finding Guide
New Mexico Ornithological Society, Albuquerque, 1984

New York
Susan Roney Drennan
Where to Find Birds in New York State: The Top 500 Sites
Syracuse University Press, Syracuse, New York, 1981

North Dakota
Kevin J. Zimmer
A Birder's Guide to North Dakota
L & P Press, Denver, 1979

Ohio
Tom Thomson
Birding in Ohio
Indiana University Press, Bloomington, Indiana, 1983

Oklahoma
Elizabeth Hayes, editor
A Guide to Birding in Oklahoma
Tulsa Audubon Society, 1985

Oregon
Fred L. Ramsey
Birding Oregon
Audubon Society of Corvallis, Oregon, 1981

Pennsylvania
David B. Freeland, editor
Where to Find Birds in Western Pennsylvania
Audubon Society of Western Pennsylvania, Pittsburgh, 1975

John J. Harding and Justin J. Harding
Birding the Delaware Valley Region
Temple University Press, Philadelphia, 1980

Tennessee
Michael Lee Bierly
Bird Finding in Tennessee
Privately published, 1980

This state-by-state list covers some of the most useful guides to finding birds. Note that some states do not have comprehensive local guides.

Texas
Edward A. Kutac
Texas Birds: Where They Are and How to Find Them
Lone Star Books, Houston, 1982

James A. Lane and Harold R. Holt
A Birder's Guide to the Rio Grande Valley of Texas
L & P Press, Denver, revised edition, 1986

James A. Lane, John Tveten, and Harold R. Holt
A Birder's Guide to the Texas Coast
L & P Press, Denver, revised edition, 1984

Utah
William H. Behle and Michael L. Perry
Utah Birds: Check-list, Seasonal and Ecological Occurrence Charts and Guides to Bird Finding
Utah Museum of Natural History, Salt Lake City, 1975

Vermont
Walter G. Ellison
Guide to Bird Finding in Vermont
Vermont Institute of Natural Science, Woodstock, Vermont, 1983

Virginia
Claudia Wilds
Finding Birds in the National Capital Area
Smithsonian Institution Press, Washington, D.C., 1983

Washington
Eugene S. Hunn
Birding in Seattle and King County
Pacific Search Press, Seattle, 1982

Terence R. Wahl and Dennis R. Paulson
Guide to Bird Finding in Washington
Privately published, Bellingham, Washington, revised edition, 1986

Wisconsin
Darryl D. Tessen
Wisconsin's Favorite Bird Haunts
Wisconsin Society for Ornithology, Green Bay, Wisconsin, 1976

Wisconsin's Favorite Bird Haunts: A Supplement to the 1976 Edition
Wisconsin Society for Ornithology, Green Bay, Wisconsin, 1979

Bird Groups
Albatrosses to Nightjars

The 62 bird groups described here are informal groups whose members clearly resemble one another in behavior, shape and posture, or color and pattern. These groups do not always correspond to families and subfamilies. For easy reference, the descriptions are brief and concentrate on field marks shared by all members of a group.

Albatrosses Very large seabirds with long, narrow, pointed wings. Often soar in stiff-winged flight, with slow wingbeats; swim and feed at surface.

Anis Black relatives of cuckoos with flattened, parrotlike bills and long, rounded tails. Flap and coast in loose-winged flight.

Auks Stocky seabirds with large heads and short wings; some have facial plumes; bill variable, and parrotlike in puffins. Posture often vertical. Swim and dive from surface; fly with rapid wingbeats.

• **Blackbirds** Mainly black, sparrow- to pigeon-sized birds that walk on ground, fly in bunches, and often mob predators.

• **Bluebirds** Sparrow-sized thrushes with blue plumage and slender bills. Often perch with tail pointed downward.

Boobies and Gannets Goose-sized to very large seabirds with stout, spear-shaped bills and long, narrow, pointed wings. Fly in beeline with slow wingbeats; dive for fish from air.

• **Buntings** Brightly or boldly colored birds with short, conical bills; females greenish or brown.

• **Chickadees** Sparrow-sized birds with dark crowns, black throats, and white cheek patches. Forage rapidly in foliage and often hang upside down from twigs.

Chickenlike Birds Stocky, ground-dwelling birds with small heads; often with crest, head or neck plumes, or combs; bill chickenlike; wings broad and rounded. Walk or run; flush with rapid wingbeats. The very large Wild Turkey belongs to this group.

Cormorants Dark diving birds with slender, hooked bills. Swim with body low. Stand upright; hold wings open to dry after diving. Fly in a line or a V, with long neck extended, flapping and coasting.

• **Crows and Ravens** Large black birds with stout bills; tail wedge-shaped in ravens. Walk on ground; often travel in bunches and mob predators; ravens soar.

Cuckoos Slender, robin-sized birds with slightly downcurved bills and long tails. Skulk in vegetation; fly very fast in beeline.

Ducks Stocky or slender waterfowl, smaller than geese or swans, with flattened bills. Dabbling ducks feed at surface, often tipping up; bay and sea ducks dive from surface. Most fly in formations, on rapid wingbeats; some fly with neck extended or have erratic flight.

• **Finches** Seed-eating birds; usually with red, pink, or yellow; most have short, conical bills, but crossbills have crossed mandibles for extracting seeds from cones. Fly in bunches, with bounding flight.

• **Flycatchers** Mostly dully colored, very small to robin-sized birds. Most perch upright and dart out after passing insects; many flick tail; hold tail pointed downward. (Strictly speaking not songbirds, but so closely related in behavior, shape and posture, and color and pattern that the group is included here as songbirds.)

Geese Long-necked waterfowl with stout bills. Swim and feed at water surface, often tipping up; usually walk on land. Fly in formation with slow wingbeats and necks extended.

These bird groups appear in the charts at the end of each chapter. Members of each group are listed in the Group Index, with the page number of the companion field guides, AN AUDUBON HANDBOOK: EASTERN BIRDS and AN AUDUBON HANDBOOK: WESTERN BIRDS. If a chart leads you to a single species, refer directly to the guides.

• Although you do not need to know whether a bird is a songbird to identify the species, it is helpful to be familiar with this term used by birders. A dot in front of a group name indicates that its members are songbirds.

• **Gnatcatchers** Very small, slender gray birds with long tails and short, slender bills. Forage rapidly, flicking tail, and cocking tail over back.

Grebes Robin- to crow-sized diving birds that somewhat resemble ducks; some have face or head plumes; most have slender, spear-shaped bill. Seldom fly; often swim with body low in water and dive from surface.

• **Grosbeaks** Sparrow- to robin-sized birds with stout, conical bills. Males brightly colored, females brown or greenish.

Gulls Mainly white water birds with stout, very slightly hooked bills, long, narrow, pointed wings, and usually short, fan-shaped tails. Swim and feed at water surface; mob predators; rob food from other birds; larger species soar.

Hummingbirds Very small to sparrow-sized nectar-feeders with very long and needlelike bills. Usually hover at flowers with very rapid wingbeats and fly in a beeline.

Jaegers and Skuas Dark seabirds with bill gull-like but more hooked; wings long, narrow, pointed. Jaegers have long central tail feathers; skuas have short, fan-shaped tails. Flight erratic, but graceful and swallowlike in smaller species; all rob food from other birds.

• **Jays** Mainly blue birds with stout bills and rounded tails; some have pointed crests. Flight bounding; mob predators.

• **Juncos** Small birds with short, conical bills and white outer tail feathers. Flush when disturbed. Dark-eyed Junco hops; Yellow-eyed Junco walks.

Kingfishers Large-headed birds with bushy crests and stout, spear-shaped bills. Perch or hover over water, then dive to capture fish.

• **Kinglets** Very small, slender-billed birds; olive-green with a bright crown patch. Often flick tails; forage rapidly in foliage, hover, and hang upside down.

• **Larks** Streaked birds with slender bills and white outer tail feathers. Walk or run on ground, flush when disturbed; have bounding flight, and usually fly in bunches.

Long-legged Waders Pigeon-sized to very large wading birds with long necks and long legs. Herons, egrets, and bitterns fly with neck folded over back; others fly with neck extended.

• **Longspurs** Sparrow-sized birds with short, conical bills. Walk on ground; flush when disturbed. Fly in bunches, with bounding flight and simultaneous banking.

Loons Goose-sized or very large water birds with stout, spear-shaped bills and very short tails. Dive from water surface; often swim with body low. Fly in a beeline, with long neck extended.

• **Magpies** Black-and-white birds with stout bills and long, pointed tails. Walk on ground.

• **Meadowlarks** Brown, open-country birds with long, pointed bills. Walk on ground and flush when disturbed.

Nightjars Slender, aerial birds, with large heads, that perch in horizontal posture and catch insects at night or dusk. Flight mothlike (true nightjars) or loose-winged and erratic (nighthawks). Clad in "dead-leaf" patterns; flush from ground.

Bird Groups
Nuthatches to Wrens

Twenty species are so distinctive in behavior or shape and posture that they form a "group" of one. These are not included here, because you will find them easily by name in a guide index.

- **Nuthatches** Stocky, short-necked birds with long, pointed bills and short tails. Scramble up or down trees in search of insects.

- **Orioles** Black-and-orange or black-and-yellow birds that forage slowly in trees, sometimes feed at flowers.

Owls Stocky, mainly nocturnal birds of prey with vertical posture; large, rounded heads, often with ear tufts. Most have camouflage colors. A few hunt during the day.

Pelicans Very large water birds with huge bills. Fly in formations with neck folded; soar, flap, and coast, with slow wingbeats.

Pigeons and Doves Small-headed birds with fan-shaped or pointed tails. Walk on ground; fly very fast in beeline, with rapid wingbeats; often fly in bunches, sometimes with simultaneous banking.

- **Pipits** Sparrow-sized, streaked birds with short, slender bills and white outer tail feathers, usually seen in open country or alpine tundra. Walk or run on ground, flush, and often fly in bunches.

Rails Brown-streaked marsh birds with small head; bill long and slim or chickenlike; wings short, broad, and rounded; tail very short. Walk, run, or wade; probe in mud; flush from grass and fly with rapid wingbeats.

Raptors Hook-billed, predatory birds, often with vertical posture. Head naked in vultures. Most raptors have long, broad, rounded wings, but wings narrow and pointed in falcons and some kites; tail usually fan-shaped or long and rounded. Often soar or flap and coast, some with wings in dihedral.

Shearwaters Seabirds with long, narrow, pointed wings and short, fan-shaped tails. Flap and coast in stiff-winged flight, often in bunches; swim and feed at surface.

Shorebirds A varied group of birds usually seen on or near water; they have long, narrow, pointed wings; larger species have long necks; bill usually short and slender, but in some downcurved, upcurved, or long and straight. Walk or run, and often wade; a few bob head or tail; fly with rapid wingbeats, often in bunches or formations, with simultaneous banking. A few species live chiefly on land.

- **Shrikes** Large-headed, predatory birds with horizontal posture and stout, hooked bill; gray with black mask. Swoop up to perch when landing.

- **Sparrows** Brown-streaked birds with short, conical bills. Skulk in vegetation; forage on ground and flush when disturbed; some scratch noisily in leaves.

Storm-petrels Blackish or gray sparrow- to robin-sized seabirds; tail squared, shallowly forked, or wedge-shaped; some have white rump. Flight swallowlike; hover at water surface to feed.

- **Swallows** Highly aerial sparrow- to robin-sized birds with long, narrow, pointed wings and very small bill. Flight graceful, erratic, and buoyant.

Swans Very large, long-necked white waterfowl with long, gooselike bills. Swim and feed at surface, often tipping up. Fly in formation with slow wingbeats, neck extended.

Anhinga
Budgerigar
• Red-whiskered Bulbul
• Bushtit
• Gray Catbird
• Brown Creeper
• Dickcissel
• American Dipper
Magnificent Frigatebird
Northern Jacana
• Northern Mockingbird
• Crested Myna

• Clark's Nutcracker
• Phainopepla
Greater Roadrunner
Black Skimmer
• European Starling
Elegant Trogon
• Verdin
• Wrentit

Swifts Highly aerial sparrow-sized birds with long, narrow, pointed wings and very short bills. Flight erratic; wingbeats rapid. Wings held more stiffly and beat more rapidly than those of the swallows.

• **Tanagers** Sparrow- or robin-sized birds with stout but not conical bills. Usually forage in trees. Males brightly colored, females olive or yellowish.

Terns Slender water birds with horizontal posture; bill slim and pointed or stout and spear-shaped; some have bushy crest; wings long, narrow, pointed; tail usually forked. Flight swallowlike; hover over water and dive from air; mob predators.

• **Thrashers** Slender birds, mostly with downcurved bills and long tails. Walk or run on ground, scratch noisily in leaves, and skulk in thickets.

• **Thrushes** Slender-billed, brown birds that often forage on ground, flipping over leaves.

• **Titmice** Sparrow-sized gray birds with pointed crest, large head, very small bills. Forage rapidly in foliage, and hang upside down from twigs.

• **Towhees** Ground-dwelling birds with short, conical bills and long, usually rounded tails. Skulk and, when disturbed, flush; scratch noisily in leaves.

Tropicbirds Mainly white seabirds with stout, spear-shaped bills, long, narrow pointed wings, and very long central tail feathers. Dive for fish from air.

• **Vireos** Mainly olive, very small to sparrow-sized birds with short, slender bills. Forage slowly for insects in foliage. Several have white wing bars, eye-rings, or spectacles.

• **Warblers** Very small to sparrow-sized birds with short, slender bills; usually yellow, olive or blue-gray. Most forage rapidly in foliage; some hawk for insects; a few walk on ground or skulk in vegetation; some bob or flick tail; a few fan tail.

• **Waterthrushes** Ground-dwelling warblers with short, slender bills. Usually walk or run on ground near water, bobbing tail.

• **Waxwings** Crested birds clad in soft browns and grays, with yellow tips to tail feathers. Travel in flocks and have bounding flight. Feed on berries and at flowers; may hawk for insects.

Woodpeckers Mainly black-and-white birds with vertical posture and strong, chisel-like bills. Hitch up tree trunks; drum on wood; have bounding flight.

• **Wrens** Stocky, brown birds with large heads, slightly downcurved bills, and short tails. Often skulk in vegetation; hold tail over back.

Group Index
Albatrosses to Ducks

The charts of field marks at the end of each chapter refer to bird groups as well as to individual species. Generally the groups consist of categories of birds such as warblers that are used by scientists as well as birders. A few groups such as shorebirds and long-legged waders are specially keyed to our charts.

Bird groups are listed alphabetically in boldface type. Under each group name, the species included are listed by common name, along with the page numbers for the companion guides, AN AUDUBON HANDBOOK: EASTERN BIRDS and AN AUDUBON HANDBOOK: WESTERN BIRDS.

References to the eastern volume (E) are given first, followed by those from the western volume (W). The names of subgroups are italicized; for example, "Bay Ducks" is a subgroup of the larger category "Ducks." The 20 species that form "groups" of one are also included here.

References to the eastern volume (E) are given first, followed by those from the western volume (W). The names of subgroups are italicized; for example, "Bay Ducks" is a subgroup of the larger category "Ducks." The 20 species that form "groups" of one are also included here.

Group Index
Shorebirds to Tropicbirds

Group Index
Verdin to Wrentit

References to the eastern volume (E) are given first, followed by those from the western volume (W). The names of subgroups are italicized; for example, "Bay Ducks" is a subgroup of the larger category "Ducks." The 20 species that form "groups" of one are also included here.

Williamson's, 331W
Yellow-bellied, 300E, 332W

Woodpecker
Acorn, 330W
Black-backed, 304E, 337W
Downy, 302E, 334W
Gila, 325W
Golden-fronted, 306E, 324W
Hairy, 301E, 335W
Ivory-billed, 299E
Ladder-backed, 309E, 333W
Lewis', 322W
Nuttall's, 333W
Pileated, 299E, 329W
Red-bellied, 307E
Red-cockaded, 303E
Red-headed, 298E, 330W
Strickland's, 327W
Three-toed, 305E, 336W
White-headed, 328W

Wrens
Bewick's, 317E, 346W
"Brown-throated," 347W
Cactus, 315E, 344W
Canyon, 322E, 343W
Carolina, 316E
House, 320E, 347W
Marsh, 318E, 345W
Rock, 323E, 342W
Sedge, 319E, 347W
Winter, 321E, 348W

Wrentit, 379W

Parts of a Bird

The parts of a bird labeled on this drawing are important as field marks, not only singly, but in combination as patterns that are helpful in identifying species.

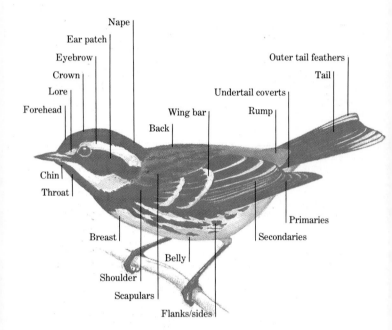

Photo Credits
A–Ke

Amwest: Rita Summers, 221(4L).
Animals Animals: George H. H. Huey, 10(1), 14.
Robert H. Armstrong: 125(4R), 128(3), 191 (4R), 223(2L), 224(3), 227(1R), 243(3R), 245(2L).
Ron Austing: 81(2), 95(1), 143(2L), 154(2), 182(4), 223(1R), 229(1R), 230(4), 233(4R), 235(1R), 237(2L), 247(2R).
Stephen F. Bailey: 149(4L).
Kathleen Blanchard: 29(2L).
William J. Bolte: 38(4R), 39(3R).
L. Page Brown: 66(2R), 136(1), 140(2).
Fred Bruemmer: 148(1), 193(4R).
Franz J. Camenzind: 278(4).
S.R. Cannings: 229(3L).
Ken Carmichael: 153(4R), 244(4).
N.R. Christensen: 151(1L), 152(3), 183(3R), 185(2L), 191(3L), 195(3L).
Herbert Clarke: 13(2L), 35(1), 38(1L), 52(1L), 58(1L, 2L), 143(3L), 153(1L), 195(4L), 224(2), 225(2R), 235(3L), 238(3), 239(1R), 241(1R, 4L).
Bruce Coleman, Inc.: Keith Gunnar, 11(1L); Mike Price, 146(2).
Cornell Laboratory of Ornithology: Hal Brown, 231(4R); John S. Dunning, 225(1R), 229(1L), 235(4R); Bill Dyer, 225(4L), 228(2), 281(1L); B.B. Hall, 280(2); Michael Hopiak, 232(1), 235(2R), 236(2), 244(3); William A. Paff, 69(2L); O.S. Pettingill, 225(2L); Mary Tremaine, 66(2L), 227(3L).
Betty Darling Cottrille: 43(2L), 45(1R, 2R), 49(1), 51(2L), 61(1), 235(2L), 245(1L).
Dorothy W. Crumb: 149(4R).
Kent and Donna Dannen: 11(1R), 12(4), 13(4L,4R), 57(1L,1R), 76(2L), 138(1), 225(3L,4R), 237(1R), 240(1), 281(2L), 282.
Harry N. Darrow: 133(3R), 138(2), 139(2L, 2R), 144(4), 145(2L), 146(1, 4), 149(2L), 150(2), 151(1R, 2L, 2R), 193(1L,2L).
Jack Dermid: 43(2R), 101(1), 124(1), 126(1), 127(1L, 4L,4R), 129(3L), 131(4R), 139(4R), 140(3),

144(2), 145(2R), 188(4), 220(2), 230(1), 240(2), 247(3R).
Adrian J. Dignan: 10(2), 18, 30(1), 55(1), 140(4), 247(3L), 277(3R, 4L).
Larry R. Ditto: 38(3R), 39(3L), 44(2R), 55(2), 77(1R), 80(2), 85(2), 132(3), 135(1L), 140(1), 141(2R), 147(1L, 3L), 148(4), 150(3), 152(2), 191(3L), 281(4L).
Harry Engels: 13(1R), 95(2), 125(3R), 129(2L), 182(2), 187(3L), 188(3), 189(2L), 195(3L), 221(1L), 228(3).
Diane Ensign: 28(2L), 76(1R).
Entheos/ Steven C. Wilson: 8, 24, 28(2R), 29(1L), 31(1), 40(1), 42(1L), 56(2R), 64(4), 65(1), 66(1R, 3L), 67(3R), 68(1L), 69(1L, 2R), 74, 76(1L), 77(1L).
Jon Farrar: 39(4R), 93(1), 124(3), 125(2L), 141(4L), 186(1), 187(2R).
Jacob Faust: 155(2L).
Tim Fitzharris: 11(4L), 13(1L), 67(4R), 98(1), 124(4), 125(3L), 155(4L), 182(1), 187(4L), 188(2), 189(3L,4R), 223(4R).
Jeff Foott: 26(2), 66(4L), 84(2), 86(2), 125(1R), 133(4R), 189(2R), 220(3).
Thomas W. French: 28(1R).
Susan Gibler: 64(3).
François Gohier: 10(4), 38(3L), 67(2L), 120, 148(3), 149(3R), 184(1).
James M. Greaves: 141(1L), 148(2), 234(4), 281(1R).
William E. Grenfell, Jr.: 145(4L), 221(3R).
Joseph A. Grzybowski: 136(3), 137(3R).
F. Eugene Hester: 45(2L), 154(3).
Irene Hinke-Sacilotto: 107(2), 182(3), 183(1R,3L), 188(1), 223(1L).
Gord James: 242(2), 245(4R).
Isidor Jeklin: 106(1), 129(1L,1R), 155(3L,3R), 185(3L), 194(2), 229(2L), 239(3R), 243(2L).
David B. Johnson: 128(4).
Steven C. Kaufman: 152(4), 183(2R), 184(2), 276(1).
G.C. Kelley: 10(3), 12(3), 38(4L),

Photo Credits
Ki–Z

41(2), 44(2L), 59(2R), 84(1), 85(1), 86(1), 87(1, 2), 91(2), 92(2), 93(2), 100(2), 103(2), 124(2), 125(2R,4L), 126(4), 127(3R), 128(1), 130(4), 131(4L), 141(4R), 147(2L,4L), 151(3L,4L), 153(4L), 155(2R), 183(1L), 184(3), 187(1L,1R), 189(3R), 191(2L), 192(3), 193(3R), 194(1), 224(4), 226(1), 230(3), 232(3), 244(2), 272, 279(4L), 280(1), 281(2R,4R).
Helen Kittinger: 39(4L), 238(4).
Dwight Kuhn: 139(1R).
Peter La Tourrette: 82(1), 141(3L), 142(2, 3), 143(1R,4L), 185(3R), 186(4), 221(2R), 228(4), 231(1R), 235(4L), 236(3, 4), 239(4R), 242(4), 243(3L,4R).
Wayne Lankinen: 13(3L), 133(3L), 137(1R), 138(4), 139(4L), 145(3L,3R), 191(3R), 195(2R), 221(2L), 223(3L), 227(2R), 233(2R), 238(1), 239(4L), 246(3), 247(2L,4R), 249(2R), 278(1, 2).
Frans Lanting: 59(1L), 61(2), 131(1L), 143(2R,4R).
Calvin Larsen: 40(4).
Tom and Pat Leeson: 80(1), 125(1L).
John A. Lynch: 30(2), 40(3), 42(1R), 50(2L), 51(1R).
Tom Mangelsen: 59(2L), 82(2), 90(1), 94(2), 139(3R), 149(1R), 195(2L).
Thomas W. Martin: 143(3R), 191(1L,1R), 192(2), 195(1R), 230(2), 233(2L), 236(1), 237(2R), 242(1), 276(3), 279(1L,3L, 3R).
Joe McDonald: 185(2R), 243(4L).
Wyman P. Meinzer, Jr.: 135(1R), 149(1L), 151(4R), 153(3R).
Anthony Mercieca: 106(2), 185(4R), 222(1, 4), 224(1), 226(4), 229(4R), 231(3L,3R), 233(3R), 239(2L,3L), 243(2R), 249(1L), 251(1L).
C. Allan Morgan: 51(1L), 64(2), 66(1L,3R), 89(2), 133(1R), 135(4R), 154(4), 243(1L).
Arthur Morris: 70(1), 76(2R), 126(2, 3) 127(2L,2R), 129(3R), 130(2), 131(2L), 134(3), 135(3L,3R),

141(3R), 150(1), 153(3L), 184(4), 193(3L), 227(3R).
Blair Nikula: 222(2).
Jerry R. Oldenettel: 134(4), 135(4L), 189(1L), 237(4L).
James F. Parnell: 44(1L,1R), 52(2R), 53(1R,2L), 146(3), 147(4R).
C.W. Perkins: 221(3L).
Wayne R. Petersen: 141(1R).
John C. Pitcher: 81(1).
Rod Planck: 12(1), 102(2), 178, 190(4), 229(4L), 237(4R).
Robert Potts: 11(2R), 67(1L, 2R).
H. Douglas Pratt: 68(1R).
Betty Randall: 26(1), 48(2), 56(1R), 57(2R), 60(1), 67(1R, 3L), 70(2), 71(1, 2), 185(4L), 239(1L), 250(1), 279(2R).
C. Gable Ray: 28(1L), 29(1R), 32(1), 34(2), 39(1L), 41(4), 54(2), 60(2), 65(2, 3, 4), 66(4R), 189(4L), 241(2L).
Laura Riley: 89(1), 133(2R), 139(3L).
Peter M. Roberts: 147(1R).
Root Resources: Ben Goldstein, 104(1); Stan Osolinski, 185(1L).
Kenneth V. Rosenberg: 45(1L), 50(1L,1R), 134(1).
Leonard Lee Rue III: 105(2), 221(1R), 223(2R).
Esther Schmidt: 129(2R).
Charles and Elizabeth Schwartz: 104(2).
Ray Schwartz: 50(2R), 130(1), 131(1R).
John Shaw: 83(1), 102(1), 103(1), 147(2R), 183(2L), 190(1), 229(3R), 276(2), 279(1R).
Ervio Sian: 234(2).
Perry D. Slocum: 231(2R), 244(1).
Arnold Small: 187(2L), 189(1R), 223(3R), 231(4L), 232(2), 234(1), 235(3R), 240(4), 241(4R), 246(1), 277(2L), 281(3R).
Tom Stack & Associates: T.J. Cawley, 233(1R); John Gerlach, 220(1), 222(3); G.C. Kelley, 142(4); Mark Newman, 144(1), 145(1R); Rod Planck, 131(3L), 243(1R), 248(1), 278(3); Milton Rand,

Index to the Charts
Albatrosses to Grebes

The bird groups and species listed here appear on the charts of field marks at the end of each chapter. Page numbers refer to each double-page chart.

*For descriptions of the groups,
refer to Bird Groups, page 294.
To learn members of each group,
check the Group Index, page 298.*

The bird groups and species listed here appear on the charts of field marks at the end of each chapter. Page numbers refer to each double-page chart.

Index to the Charts
Quail to Warbler

The bird groups and species listed here appear on the charts of field marks at the end of each chapter. Page numbers refer to each double-page chart.

Index to the Charts
Warbler to Yellowthroat

Chanticleer Staff

Publisher: Paul Steiner
Editor-in-Chief: Gudrun Buettner
Executive Editor: Susan Costello
Managing Editor: Jane Opper
Associate Editor: David Allen
Assistant Editor: Eva Colin
Production Manager: Helga Lose
Production Assistants: Gina Stead-Thomas, Philip Rappaport
Art Director: Carol Nehring
Art Associate: Ayn Svoboda
Art Assistant: Cheryl Miller
Picture Library: Edward Douglas
Drawings: Paul Singer
Maps: Tibor G. Toth
Design: Massimo Vignelli

Companion Volumes
Audubon Handbooks

Eastern Birds
This compact guide covers every breeding species east of the Rockies, with 1,354 color photographs and 179 drawings. Picture-and-text accounts use the easy-to-remember field marks explained in *How to Identify Birds*. Twelve special pages of flight comparison photographs for swans, geese, ducks, raptors, and gulls.

Western Birds
Covering every breeding species west of the Rockies, this practical guide features 1,314 color photographs and 168 drawings. Picture-and-text accounts use the easy-to-remember field marks explained in *How to Identify Birds*. Twelve special pages of flight comparison photographs for swans, geese, ducks, raptors, and gulls.